THE LEGO® BUILDER'S HANDBOOK

THE LEGO® BUILDER'S HANDBOOK

BECOME A MASTER BUILDER

BY DEEPAK SHENOY

no starch
press®

San Francisco

Printed in China

First printing

28 27 26 25 24   1 2 3 4 5

ISBN-13: 978-1-7185-0380-9 (print)
ISBN-13: 978-1-7185-0381-6 (ebook)

 Published by No Starch Press®, Inc.
245 8th Street, San Francisco, CA 94103
phone: +1.415.863.9900
www.nostarch.com; info@nostarch.com

Publisher: William Pollock
Managing Editor: Jill Franklin
Production Manager: Sabrina Plomitallo-González
Production Editor: Jennifer Kepler
Developmental Editor: Nathan Heidelberger
Cover: Sabrina Plomitallo-González and Deepak Shenoy
Interior Design: Sabrina Plomitallo-González
Technical Reviewer: Graham E. Hancock
Copyeditor: Scout Festa
Proofreader: Katrina Horlbeck Olsen

Library of Congress Cataloging-in-Publication Data

Names: Shenoy, Deepak, 1972- author.
Title: The LEGO builder's handbook : become a master builder / by Deepak
    Shenoy.
Description: San Francisco, CA : No Starch Press, [2025] | Includes index.
Identifiers: LCCN 2024007813 (print) | LCCN 2024007814 (ebook) | ISBN
    9781718503809 (hardcover) | ISBN 9781718503816 (ebook)
Subjects: LCSH: LEGO toys.
Classification: LCC TS2301.T7 S525 2025 (print) | LCC TS2301.T7 (ebook) |
    DDC 688.7/25--dc23/eng/20240328
LC record available at https://lccn.loc.gov/2024007813
LC ebook record available at https://lccn.loc.gov/2024007814

For customer service inquiries, please contact info@nostarch.com. For information on distribution, bulk sales, corporate sales, or translations: sales@nostarch.com. For permission to translate this work: rights@nostarch.com. To report counterfeit copies or piracy: counterfeit@nostarch.com.

[DC]

For my daughter, Riya

ABOUT THE AUTHOR

Deepak Shenoy discovered the joys of LEGO later in life, while playing with his daughter. He has since become an active member of the LEGO community, sharing his creations with other LEGO fans at conventions and online through social media. His builds have been featured in print in *BrickJournal* and *Blocks*, and online through sites like Brothers Brick and BrickNerd. Shenoy lives in Pennsylvania with his wife and daughter. He works as an engineer by day and spends his spare time on various hobbies, including hiking, photography, and, of course, designing and building LEGO models.

ABOUT THE TECHNICAL REVIEWER

Graham E. Hancock is a LEGO expert who can't remember life without LEGO bricks. He's the author of several LEGO books and the editor of *Blocks*, the monthly magazine for LEGO fans. He delivers LEGO talks at events and facilitates workshops with companies around the world. He lives with more LEGO models than he has room for in London, England.

BRIEF CONTENTS

CONTENTS IN DETAIL

xiv

FOREWORD

I first started building LEGO creations a long time ago with the first LEGO X-wing set, which was released in 1998. From there, I continued to build, and after a couple of years I attended my first LEGO fan convention in 2001. Five years later, I started *BrickJournal*, a LEGO fan magazine. It's been over 25 years, and I'm still building.

My first contact with Deepak Shenoy was online. I wanted to feature his work in *BrickJournal* after seeing renders of his skyscraper models and reading his blog posts about LEGO building techniques. He wrote an article for the magazine, and then I got to meet him and see his built skyscrapers in person at BrickFair Virginia.

Some time later, Deepak reached out saying he'd written a book and asked me about writing a foreword. He sent me a draft, and let me tell you, this book is something I wish I had when I started building all those years ago. The book is a toolbox for a hobby that uses only one real tool: a brick separator.

Deepak has brought to light the unseen tools used in the LEGO hobby: a wide array of building techniques. What are the strongest ways to build a model? How do you create detail? How do you ensure a model is built accurately to scale? How do you make a curve? You can do what I did and learn through years of trial and error, or you can read this book. One method will take a lot less time!

To be clear, this isn't a book of instructions to build specific models. It's a book that will give you the tools to build what *you* want to build. If LEGO building was a class, this would be the textbook. But it's also more.

The LEGO Builder's Handbook is a reference and a guide for those wanting to start building or explore new areas of the hobby. It may inspire you to try a mosaic or sculpture or a skyscraper. There's a lot I know from experience that's in this book. But there's also a lot inside that I didn't know, even after all these years.

So, start reading and building and posting your creations online. Maybe one day you'll see a note from me asking if you'd be interested in being featured in my magazine. Just like I first contacted Deepak about his skyscrapers.

Joe Meno
Founder and editor of *BrickJournal*

ACKNOWLEDGMENTS

I would like to express my deepest gratitude to the following people, without whose contributions this book would not have been possible:

Nathan Heidelberger, my developmental editor, who worked patiently and diligently to polish and shape my manuscript into a far better book than I could have created on my own. It was a great pleasure and privilege working with him on this book.

Bill Pollock, Jill Franklin, Jennifer Kepler, Morgan Vega Gomez, and the rest of the team at No Starch Press for all their support behind the scenes.

Joe Meno for writing the foreword for the book, which is such an honor.

Graham Hancock for providing invaluable feedback during his technical review of the book.

Dave Schefcik and Jonathan Lopes for all their support and for reviewing my early drafts. Their suggestions helped make this a stronger book.

Tom Alphin, Tiago Catarino, Jeff Friesen, and Bjorn Ramant for providing early feedback on the proposed contents of the book.

Sean Kenney and the late Arthur Gugick, whose models of the Empire State Building and the Taj Mahal, respectively, opened my eyes to what LEGO was capable of and got me to dive headlong into this hobby.

Chris Doyle, Didier Enjary, Eric Harshbarger, Bruce Lowell, Holger Matthes, Bill Ward, Paul Wellington, and all the others who take the time to share their knowledge and experience with the LEGO community and provide platforms for other builders to do the same.

Rocco Buttliere, Alice Finch, Paul Hetherington, Felix Jaensch, Ryan McNaught, Anuradha Pehrson, Luca Petraglia, Zachary Steinman, Pete Strege, and Timofey Tkachev, to name just a few of the builders who continue to push the envelope with their LEGO builds and inspire me with each new build that they create.

Greg DiNapoli for being a great "internet friend" and a sounding board for my ideas, no matter how crazy they may sound (like the one I had about writing a book someday).

Last but not least, my wife, Roopa, and daughter, Riya, for all their support during the process of writing this book.

INTRODUCTION

LEGO needs no introduction. For many people, the globally recognized brand is synonymous with the colorful plastic bricks they grew up playing with. LEGO has indeed been a popular children's toy for generations. At the same time, though, it's also transcended its origins as a toy and become a medium of choice for adults looking to explore their creative side or to simply unwind from their daily routines.

Today, AFOLs (short for *adult fans of LEGO*) make up a vibrant and growing worldwide community. They share their ideas and creations online through social media, as well as in person at LEGO conventions. These conventions feature elaborate displays of MOCs (or *my own creations*, a term for original models designed and built by LEGO fans) and can draw thousands of visitors.

Whether they have rediscovered LEGO after a "dark age" (a period in their life when they lost interest in LEGO) or are encountering it for the very first time, AFOLs typically (re)start their building journey with official LEGO sets. These sets come with step-by-step building instructions and all the pieces you need to realize the model shown on the box. In the earliest days of LEGO, a set's building instructions could fit on a single sheet of paper. Over the years, however, LEGO sets have grown much more complex, especially the ones geared toward adults. Some require a thousand or more separate building steps, and the instructions come in sizable booklets—often more than one.

Official LEGO sets are chock-full of new and interesting building techniques, and you can learn a lot from studying them. But there's no reason to limit yourself to these sets. Like many AFOLs, you can take inspiration from official sets and use them as a jumping-off point for your own explorations. Think of this book as a road map for that process, a guide on your journey from following step-by-step instructions to developing original MOCs. I'll identify useful building techniques from official sets, share insights from the process of creating my own models, and spotlight some other AFOLs' ideas as well.

WHO THIS BOOK IS FOR

If you are an AFOL or TFOL (teenage fan of LEGO) who enjoys building official LEGO sets and would like to take a first step toward building something of your own design, this book is just what you need to get started. This isn't to say, however, that prior LEGO building experience is required to get the most out of the book. We'll start at a very basic level and discuss fundamental concepts related to the LEGO system, the most common types of LEGO elements, and how they fit together. Then, with each new chapter, we'll build on those concepts as I introduce other types of LEGO elements and techniques that will help you take your skills to the next level.

While you don't need prior LEGO experience, I'm assuming you have some familiarity with basic mathematics, especially angles and geometry. But I've included explanations of important concepts like the Pythagorean theorem, in case you're feeling a bit rusty. Also, the last part of the book explores LEGO builds that are designed with the help of software. Here it will be useful to have access to a computer so you can download and experiment with some of the tools mentioned in the chapters.

WHAT THIS BOOK ISN'T

This book won't tell you what to build or provide step-by-step instructions for building any specific LEGO models. Instead, by outlining generally applicable techniques, my aim is to lay the groundwork for you to develop your own designs. I hope the book will inspire you to continue your exploration of these techniques through trial and error or by tapping into the wealth of other LEGO resources that are available in print and online.

WHAT YOU'LL LEARN

Here's a quick overview of what you'll learn from this book's chapters.

Part I: The Basics will bring you up to speed on the general concepts and simple building techniques used throughout the book so you can start to design your own creations.

CHAPTER 1: THE LEGO SYSTEM Provides an overview of the toy's history and the inherent traits that make the LEGO system click. It introduces the most common types of LEGO elements, as well as their geometry, and the measurement units you'll use as a LEGO builder. This chapter also weighs the pros and cons of physical versus digital building.

CHAPTER 2: ALL ABOUT SCALE Presents the concepts of scale and proportion and how they come into play for LEGO builds. In this chapter, you'll also learn how to design realistic LEGO replicas of recognizable real-world structures.

CHAPTER 3: BASIC LEGO TECHNIQUES Covers simple building techniques borrowed from masonry that will be indispensable in your LEGO builds. You'll explore different bond patterns and how to ensure that your models hold together well and are as sturdy as possible. You'll also review different types of LEGO slope pieces and how they can be used.

Part II: Breaking Free from the Grid will demonstrate some powerful techniques for breaking free from the confines of the square LEGO grid. You'll learn about techniques to offset bricks by fractions of the grid unit, build sideways, create angled walls, and even approximate round shapes.

CHAPTER 4: HALF-STUD OFFSETS Outlines the different types of jumper plates and how you can use offset techniques to make your models more realistic. These techniques will allow you to add surface texture, create recessed windows or wall sections, and achieve smoother tapers in your LEGO models.

CHAPTER 5: SIDEWAYS BUILDING (SNOT) Shows how to use SNOT (studs not on top) techniques to attach elements sideways instead of stacking them on top of each other. You'll learn to take full advantage of the geometry and proportions of LEGO elements to create shapes and details that wouldn't be possible with simple stacking.

CHAPTER 6: ANGLED WALLS Demonstrates that you aren't always limited to 90-degree angles when building walls with LEGO. We'll cover useful LEGO elements for building angled walls, such as hinge plates and turntables, as well as tips for building them digitally in BrickLink Studio.

CHAPTER 7: ROUND SHAPES Illustrates the different ways of creating shapes like cylinders, circular walls, domes, and even spheres with regular bricks and plates, as well as with curved LEGO elements. You'll learn techniques like brick bending, hinged polygons, and Lowell spheres and explore software tools like Bram's Sphere Generator that can come in handy as you plan your LEGO builds.

Part III: Computer-Assisted Builds will cover two categories of LEGO models—mosaics and sculptures—that require careful planning. You'll look at available tools and software to automate the process.

CHAPTER 8: MOSAICS   Details the process of using software to convert photographs or paintings into LEGO mosaics. You'll learn to arrange different-colored LEGO elements in a grid to form a two-dimensional image and leverage computer programs like BrickLink Studio and LEGO Art Remix to achieve realistic, eye-catching results.

CHAPTER 9: SCULPTURES   Describes the process for converting three-dimensional models of organic shapes into LEGO sculptures. These models have shapes that can't easily be broken into basic geometric components like cubes, cylinders, or spheres. You'll learn the steps to create studs-up sculptures with software assistance and use LSculpt to design studs-out sculptures.

The book concludes with answers to some questions you may have about the next steps in your LEGO journey. You'll find tips on where to buy LEGO pieces, how best to store and organize them, where to find more information on building techniques, and how to become a more active member of the LEGO community. Throughout the book, you'll come across examples from my own models of skyscrapers and other real-world landmarks. While I may be partial to architecture-themed builds, the techniques presented here are equally applicable to other types of LEGO models.

ONLINE RESOURCES

As you read, you'll find LEGO pieces listed with their official part numbers—for example, a 2×4 brick (#3001). The pieces I mention are just a small subset of the overall catalog of LEGO elements. One of the best resources to consult regarding the available types and colors of LEGO pieces is BrickLink (*https://www.bricklink.com*), an online marketplace and subsidiary of the LEGO Group, where you can order all the LEGO pieces you need to build your models. Parts on BrickLink are listed under the same numbers I reference in the book. Official LEGO sets also have numbers (which are also searchable on BrickLink); I've included these as well when I reference a particular set.

Besides being a forum for buying and selling parts, BrickLink is the home of BrickLink Studio, the LEGO Group's recommended design software. I'll periodically include tips for how to implement a technique using Studio. If you're interested in building digitally—no physical LEGO bricks required—I recommend downloading a copy of this free program.

PART I

THE BASICS

LEGO is the perfect medium for exploring your creativity and artistic skills. But how do you move beyond the canned instructions that come with the LEGO sets you can buy off the shelf? This part of the book will get you well on your way to designing your own LEGO creations by bringing you up to speed on basic terms and concepts, the most common types of LEGO elements, important LEGO measurement units, and scale and proportion. We'll also examine some basic building techniques to help you build models that are as sturdy as possible.

THE LEGO SYSTEM

We'll start our LEGO journey with a brief overview of the toy's history. You'll see how the company that makes the colorful plastic bricks we all know and love came into existence. This will give us the context in which to consider the qualities that contribute to LEGO's enduring appeal. We'll also review some of the basic types of LEGO elements and establish important terminology, as well as weigh the pros and cons of physical versus digital building. Understanding these basics will lay the groundwork for the building techniques covered in the chapters to come.

A BRIEF HISTORY OF LEGO

The story of how LEGO came to be is a great lesson in resilience and turning adversity into opportunity. The company that would become LEGO had its humble beginnings in a small woodworking business in Billund, Denmark, that was owned by carpenter Ole Kirk Christiansen. When the demand for carpentry work dried up during the Great Depression, Christiansen made the fateful decision to branch out into building simple wooden toys that would be easier to sell. His business eventually transitioned into a toy manufacturing company, and in 1934, Ole named this new company LEGO, a nod to the Danish phrase *leg godt*, meaning "play well."

FROM WOOD TO PLASTIC

When a devastating fire destroyed the LEGO factory in 1942, Ole refused to accept defeat. Instead, he used that setback as an opportunity to rebuild the factory and make it bigger and better suited to mass-producing toys. After the end of World War II, when it became harder to source wood, Ole was quick to adapt to the new trend of making toys out of plastic. He took a big risk and invested in an expensive injection molding machine. In 1949, LEGO released its first set of plastic "automatic binding bricks," setting the stage for a new generation of toys.

Ole's son Godtfred Kirk Christiansen, who would inherit the family business, was initially skeptical about the transition to plastic, but he quickly came on board with the idea once he recognized the product's potential. He envisioned the plastic bricks as the basic building blocks of an entire *system* of related products. Children wouldn't be limited to building the model shown on the box of the set; they could mix and match the bricks that came in the different sets to create original models that expressed their own creativity and imagination. For this to be possible, LEGO had to ensure that all its bricks were standardized, able to fit together no matter when they were bought or which set they came from.

THE MODERN BRICK

In 1958, LEGO patented the modern form of its brick design, which is still in use today. Two years later, after another fire destroyed LEGO's remaining inventory of wooden toys, Godtfred felt confident enough in his new strategy to discontinue the production of wooden toys altogether, eliminate other plastic toys from its product line, and focus solely on building sets made up of LEGO bricks. The rest, as they say, is history.

Godtfred's incredible foresight and vision set LEGO on the course to becoming what it is today—a global powerhouse and household name. LEGO is now the biggest toy company in the world, with a catalog that includes thousands of different building sets sold in branded stores around the globe as well as in regular retail stores. Beyond the traditional products, the LEGO brand now also encompasses theme parks, movies, TV shows, and video games.

HOW LEGO CLICKED

What's the secret to LEGO's success? Is it the company's savvy marketing strategy, or its decision to enter into licensing agreements with major pop culture franchises like Star Wars, Harry Potter, and Marvel? These factors may have helped, but the brand's enduring popularity wouldn't have been possible without some inherent traits of the LEGO system itself that were key to making the brand what it is today. Let's consider some of these traits.

SYSTEM IN PLAY

Godtfred Kirk Christiansen's vision of a "System in Play" was a new paradigm in the toy industry. According to this idea, each building set, rather than being a stand-alone toy, should be a part of a unified system. As a child grows older and their interests evolve and abilities improve, the system can grow with them and continue to provide play opportunities that engage their creativity. The LEGO pieces in one building set can be used in conjunction with pieces from any other set to build anything the child can imagine. This opens the door to new possibilities beyond the models shown on the boxes of the building sets.

BACKWARD COMPATIBILITY

Take one peek inside a LEGO store (or an online store if there isn't a brick-and-mortar one near you), and you'll immediately see that these aren't your grandfather's building blocks. The sheer number and variety of sets that are currently available can be mind-boggling. And yet, amazingly, the LEGO pieces being made today can fit together just fine with your grandfather's building blocks (if your grandfather happened to play with LEGO, that is). Even with all the changes LEGO has gone through as a company over the years, one thing it hasn't changed is the size of the brick itself. Today's LEGO bricks are fully compatible with the bricks made during the earliest days of LEGO, and they can continue to be used well into the future without any risk of them ever becoming obsolete.

TIGHT TOLERANCES

LEGO's backward compatibility and system-wide unity are ensured by the very tight tolerances enforced during the injection molding process used to shape liquified ABS (a type of thermoplastic) into LEGO bricks. The maximum allowable deviation in measurements from one brick to another is typically around 0.01 mm, thinner than a strand of hair. With such high-precision manufacturing, LEGO bricks are guaranteed to fit together perfectly no matter when they were made or in which factory.

These strict standards are part of why LEGO is such a popular medium for making scale models of buildings, ships, aircraft, and other real-world structures. Especially on larger models, even small variations in the size of the pieces can add up, causing the different sections of the model to not fit together correctly.

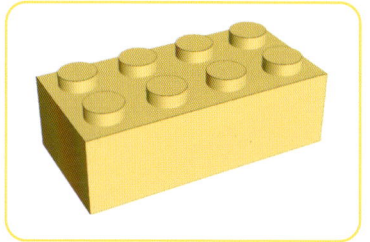

Figure 1-1: A 2×4 LEGO brick

SUPERIOR CLUTCH POWER

Clutch power is the grip that holds one LEGO piece to another. Thanks to an early design innovation from Godtfred Kirk Christiansen, LEGO bricks have an ideal amount of clutch power, making them fun and sturdy to build with but also flexible enough to reconfigure.

The first LEGO element that Godtfred patented back in 1958 was the 2×4 brick (Figure 1-1). On top, it has two rows of round bumps, or *studs*, with four studs in each row. (LEGO dimensions are generally specified in this way, telling you the number of rows and columns of studs on top of an element.) The studs are meant to connect with the underside of the brick above, allowing LEGO bricks to be stacked.

Before 1958, LEGO bricks were hollow underneath, which didn't allow for a sturdy connection between bricks (see Figure 1-2, left). Godtfred's innovation was to add three round, hollow tubes to the underside of each 2×4 brick (Figure 1-2, right). These tubes interlock perfectly with the studs of the brick below: the studs get wedged between the tubes and the sides of the brick, giving LEGO bricks much higher clutch power and stability when joined together. The space where the stud fits is known as an *anti-stud* or *stud receptacle*.

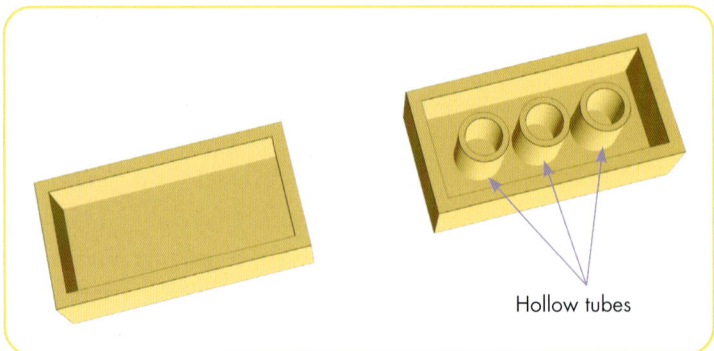

Figure 1-2: The underside of an original LEGO brick (left) and a modern LEGO brick with added hollow tubes (right)

There's a sweet spot to clutch power: it must be strong enough to allow LEGO bricks to reliably stay together, but not so strong as to make it difficult to take the bricks apart. The studs-and-tubes combination gives today's LEGO bricks just the right amount of clutch power, making it possible to assemble (and disassemble) even the most massive creations, like LEGO skyscrapers. With their durable structure, LEGO pieces can also usually be joined and taken apart again and again without significant loss of clutch power.

While it's easy to take regular bricks apart using your bare fingers, you may sometimes need a little help with types of pieces that are thinner, like plates and tiles. This is where a LEGO brick separator, shown in Figure 1-3, can come in handy.

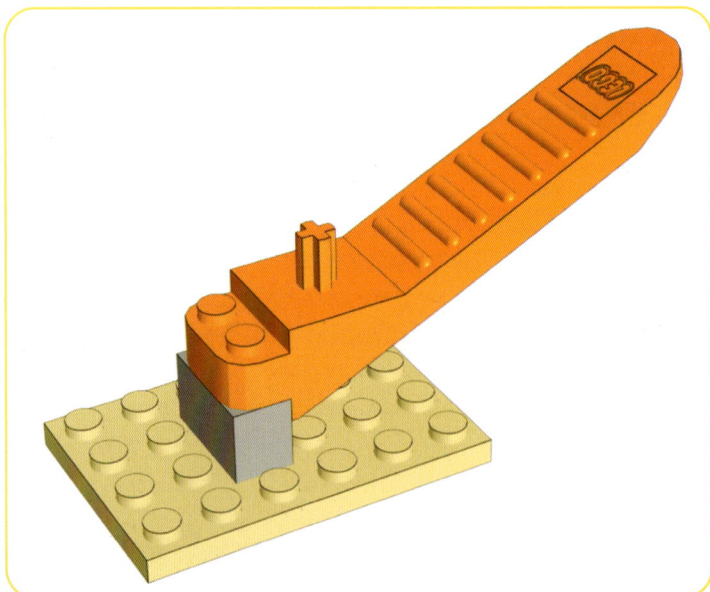

Figure 1-3: A LEGO brick separator

This plastic tool, included in some of the bigger LEGO sets, is an antidote to the legendary clutch power of LEGO elements. You can use it like a lever to quickly, and without much effort, take apart LEGO pieces that are joined together.

UNLIMITED POSSIBILITIES

The concept of interlocking bricks may seem simple, but it opens up a world of unlimited possibilities. Consider that two 2×4 bricks can be joined together in 24 different ways, as shown in Figure 1-4.

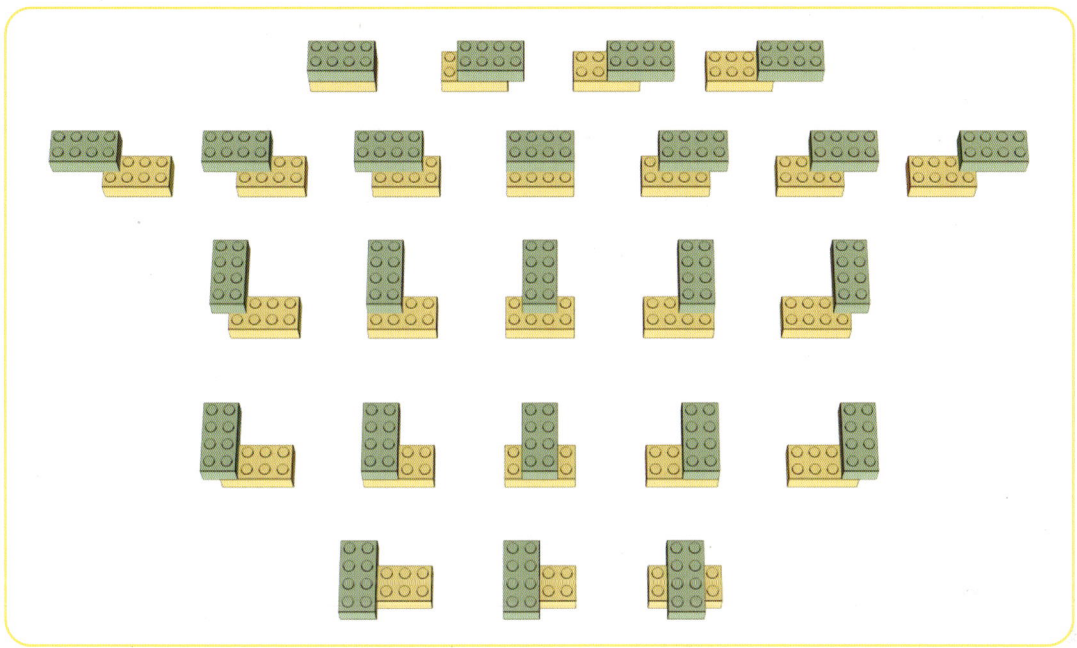

Figure 1-4: The 24 possible ways to join two 2×4 bricks

The number of possibilities goes up exponentially as you add more bricks; for example, with just six 2×4 bricks, there are 915,103,765 different combinations. Imagine the possibilities for combining all 4,000 or so different types of LEGO elements that have been produced in over 60 different colors!

ENDLESS RECONFIGURABILITY

One of the things that makes the LEGO medium so attractive to children and adults alike is the fact that your creations are never set in stone. You can always update your LEGO models months or years after originally building them to add details that you may have missed or to incorporate LEGO pieces that didn't exist before. You can also take your model apart and reuse the pieces to build something completely different.

BASIC LEGO ELEMENTS

People often refer generically to all LEGO elements as *bricks*, but technically a brick is just one specific type of element. Other common varieties of elements include plates and tiles. In this section, we'll discuss the differences between these basic types of LEGO elements and consider some other, more specialized types of elements as well.

BRICKS

Bricks are the most common type of element, the basic building blocks of the LEGO system. Most iconic of all is the 2×4 brick (shown in Figure 1-1), but bricks come in many different dimensions. A typical brick is roughly 1 cm tall and has studs on top.

The 1×1 brick, with just one stud on top, is the smallest brick available. All the bigger bricks are essentially multiples of a 1×1 brick. For instance, a 2×4 brick is the same size as two rows of four 1×1 bricks placed next to each other (see Figure 1-5).

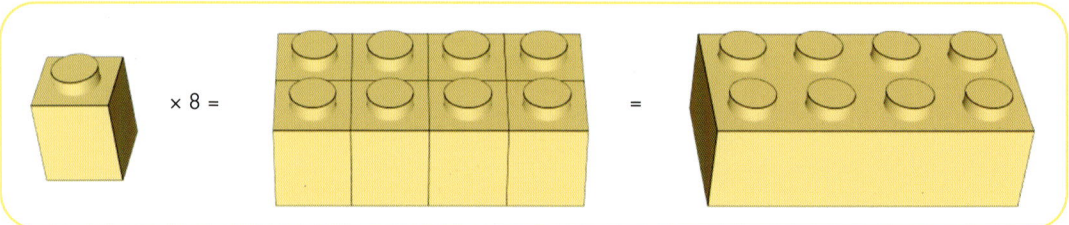

Figure 1-5: Eight 1×1 bricks arranged to have the same overall size as a 2×4 brick

Larger bricks are normally available with one or two rows of studs and an even number of studs per row—for example, 1×2, 2×2, 1×4, 2×4, 1×6, and 2×6. (There are also 1×3 and 2×3 bricks, an exception to the even-studs rule.) The longest available is the 1×16 brick.

PLATES

Plates are the thinner counterparts of bricks. The smallest is the 1×1 plate. It has the same footprint as a 1×1 brick but is only a third as tall. You therefore need a stack of three 1×1 plates to match the height of a 1×1 brick, as shown in Figure 1-6.

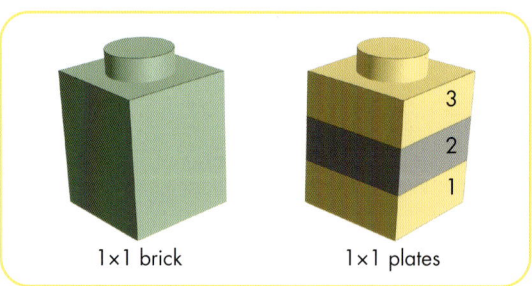

1×1 brick   1×1 plates

Figure 1-6: A stack of three 1×1 plates has the same height as a 1×1 brick.

Plates can have much bigger footprints than bricks and are generally not limited to one or two rows of studs. The 16×16 plate is the biggest one currently available in the LEGO catalog.

Baseplates are special plates that come in sizes as big as 48×48. Figure 1-7 shows a 32×32 baseplate, one of the most common sizes available.

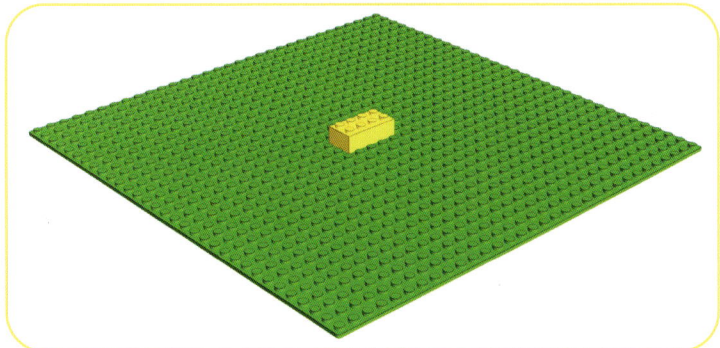

Figure 1-7: A 32×32 LEGO baseplate

Baseplates are thinner than regular plates and don't have anti-studs on their bottom, so you can't attach any elements beneath a baseplate. Baseplates are generally used as the base or foundation of a LEGO build.

TILES

Tiles have the same thickness as plates, but they don't have studs on top (see Figure 1-8). They can be used to cover up exposed studs on a model for a smooth finish.

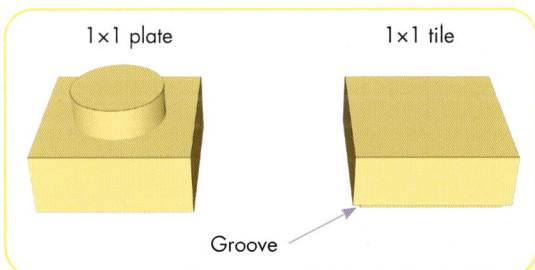

Figure 1-8: A 1×1 LEGO plate and tile

Like plates, tiles come in sizes 1×1 and up. But what do these sizes mean if tiles have no studs on top? In addition to being the name of the round bump on the top of a LEGO brick, a stud also doubles as a unit of measurement in LEGO terminology—1 stud is equivalent to the overall length or width of a 1×1 brick, plate, or tile. We'll discuss LEGO units of measurement in more detail shortly.

LEGO tiles are some of the hardest elements to detach from a model because they're so thin and smooth. Fortunately, most tiles have a little groove along their bottom edge (see Figure 1-8). This makes it easier to get a fingernail or the sharp end of a brick separator under the tile to pry it off, as shown in Figure 1-9.

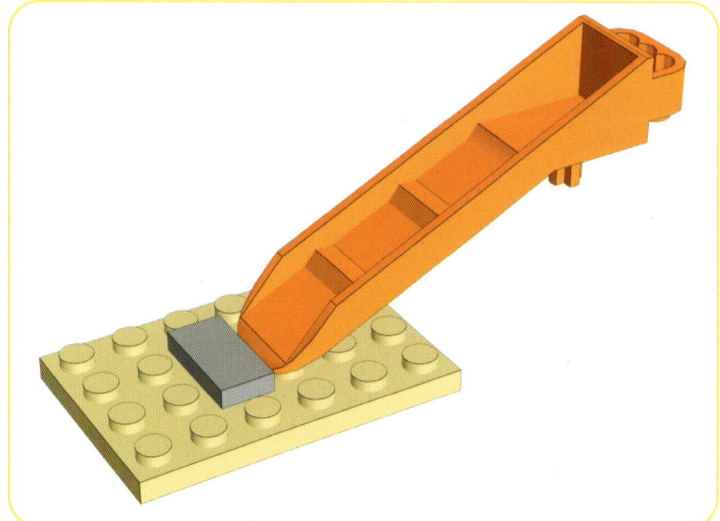

Figure 1-9: Using a brick separator to remove a LEGO tile

The smooth top surfaces of tiles make them ideal candidates for customization. You'll see that a number of LEGO sets include decorated tiles with various designs printed on their top surface, or stickers that can be affixed to the top surfaces of regular tiles.

ROUND ELEMENTS

In addition to elements with square and rectangular footprints, LEGO also makes pieces that have round footprints. Figure 1-10 shows some examples.

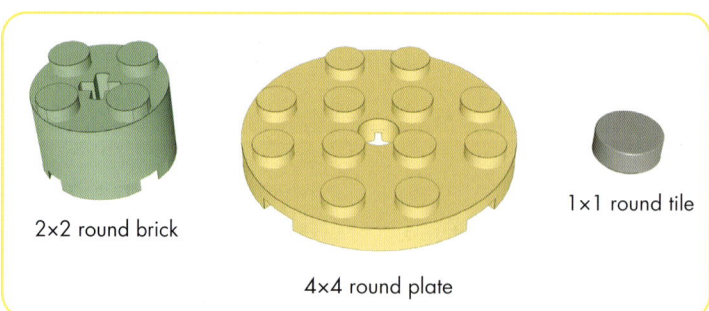

2×2 round brick

4×4 round plate

1×1 round tile

Figure 1-10: A selection of round elements

The regular square grid still determines where these round pieces can be placed—the studs on a 2×2 round brick are spaced the same as the studs on a 2×2 square brick, for example. However, because of their shape, round pieces leave gaps when placed next to each other. As a result, the applications for these elements are somewhat limited. They're good for structures like columns that can be created simply by stacking round bricks, plates, and tiles, but they can't be combined to create a bigger round shape (unlike the way rectangular bricks can be seamlessly combined to create larger rectangular shapes). We'll consider techniques for building round shapes using LEGO in Chapter 7.

TECHNIC

LEGO Technic is a line of building sets featuring moving parts and realistic mechanisms that was introduced in 1977. Technic sets include more than just regular LEGO bricks and plates; they also have specialized pieces such as beams, gears, axles, connector pins, and even electric motors. Figure 1-11 shows examples of some LEGO Technic parts.

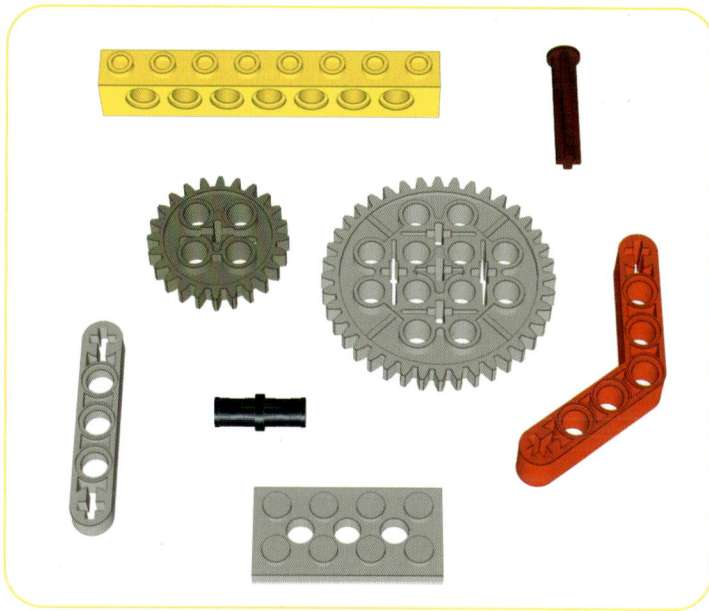

Figure 1-11: A selection of LEGO Technic parts

Early Technic sets used beams that were nothing more than regular 1×*n* bricks with circular holes in their faces. These holes could accommodate pins for connecting beams that were parallel to each other or hinged at an angle. The holes could also be used as bearings for axles so that gears and wheels could be attached for more complex mechanisms. LEGO has slowly replaced these beams with *liftarms*, which still have the holes but no longer have studs. This allows for a "studless" style of construction in more modern Technic sets. Figure 1-12 shows a comparison of these different types of Technic elements.

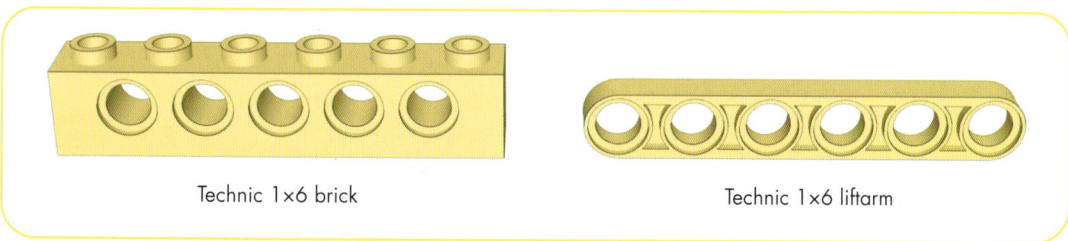

Technic 1×6 brick

Technic 1×6 liftarm

Figure 1-12: A 1×6 Technic brick and 1×6 Technic liftarm

Technic is considered separate from the main LEGO system of bricks, plates, tiles, and the like, although the two are fully compatible with each other and can be intermingled. In this book, we'll consider certain applications where it's helpful to mix Technic elements with regular LEGO system elements—for example, to help with sideways building, or SNOT (see Chapter 5). However, we won't be delving into Technic-centric building techniques. Other books cover Technic in its own right, such as *The Unofficial LEGO Technic Builder's Guide*, 2nd edition, from No Starch Press.

VARIATIONS IN STUDS AND ANTI-STUDS

Not all LEGO bricks and plates have the same kinds of studs. For example, while most studs are flat on top, Technic bricks feature *open studs*, which have a solid outer wall but are hollow in the middle. This variation stems from the details of LEGO's injection molding process, but it also has some functionality: the open studs can serve as connection points for elements like cylindrical bars, as shown in the following image. Open studs are also found on elements such as jumper plates, which we'll discuss in Chapter 4.

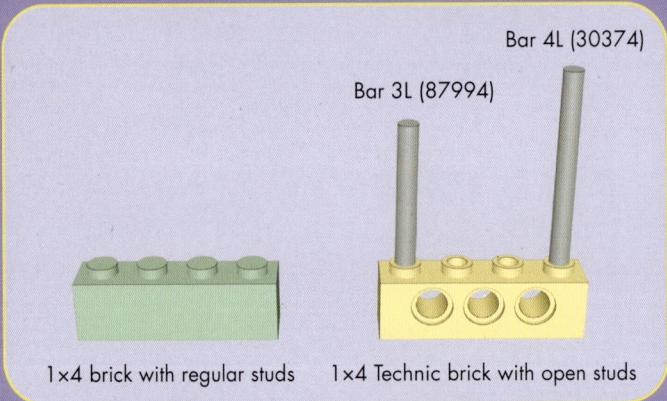

Bar 4L (30374)

Bar 3L (87994)

1×4 brick with regular studs   1×4 Technic brick with open studs

The anti-studs on the undersides of LEGO bricks also exhibit some variety. The norm is the hollow tube style illustrated in Figure 1-2, but some other designs are shown here.

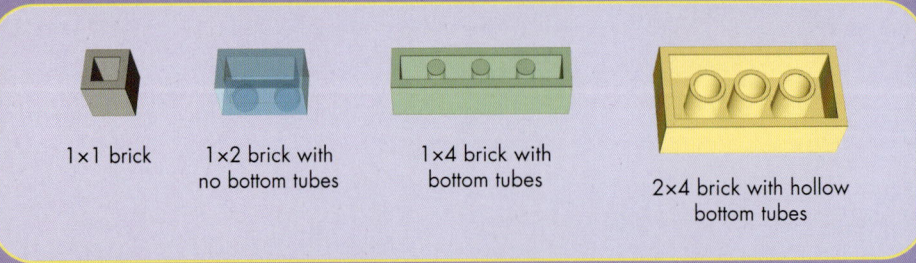

1×1 brick   1×2 brick with no bottom tubes   1×4 brick with bottom tubes   2×4 brick with hollow bottom tubes

The underside of a 1×1 brick is completely hollow because the four walls of the brick are all that are needed to create an anti-stud. Meanwhile, bricks and plates that have one row of studs (such as the 1×4 brick shown in the figure) don't have the space for hollow tubes. Instead, they have narrow, closed-off tubes separating the anti-studs. There are also bricks (especially in transparent colors) that have no bottom tubes at all. This helps keep these bricks transparent all the way through, but it comes at the expense of clutch power.

LEGO MEASUREMENT UNITS

There are quite a few different units available for measuring the dimensions of LEGO elements. You may come across LEGO measurements expressed in terms of LUs (LEGO units) or LDUs (LDraw units), to say nothing of ordinary units like centimeters. (Metric units work better for measuring LEGO than US customary units like inches.) While these sorts of units may be useful if you're working with a CAD software tool, as a builder it's usually more intuitive to think in terms of simpler units: studs and plates.

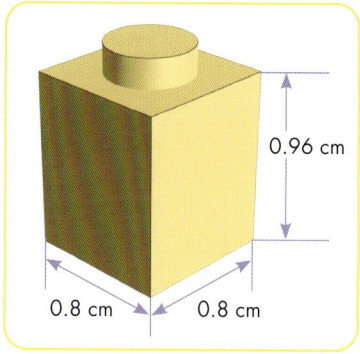

Figure 1-13: The height and width of a 1×1 brick

THE STUD

As mentioned earlier, a stud is the width of a 1×1 brick. A stud is also the distance from the center of one bump on a LEGO brick or plate to the center of the adjacent bump—a span known as the *stud pitch*. Since it defines the spacing between studs (the physical bumps), you can also think of the stud (the unit of measure) as the basic unit in the regular square grid of locations where LEGO pieces can be placed. In metric units, a stud is equivalent to 0.8 cm, although the actual width of a 1×1 brick is more like 0.78 cm to allow some clearance between bricks when they're placed directly next to each other.

If you look closely at a 1×1 brick, you'll see that it isn't a perfect cube (even if you disregard the extra height added by the stud itself). It's slightly taller than it is wide. As shown in Figure 1-13, a 1×1 brick is 0.8 cm (1 stud) wide, but 0.96 cm (1.2 studs) tall.

THE PLATE

A *plate* is a unit of measure equivalent to the height of a LEGO plate (ignoring the stud on top). As you've seen, a plate is a third as tall as a brick, so 1 plate is equal to 0.96/3 = 0.32 cm. Thinking in terms of plates, a 1×1 brick is 3 plates tall and 0.8/0.32 = 2.5 plates wide. These dimensions are illustrated in Figure 1-14.

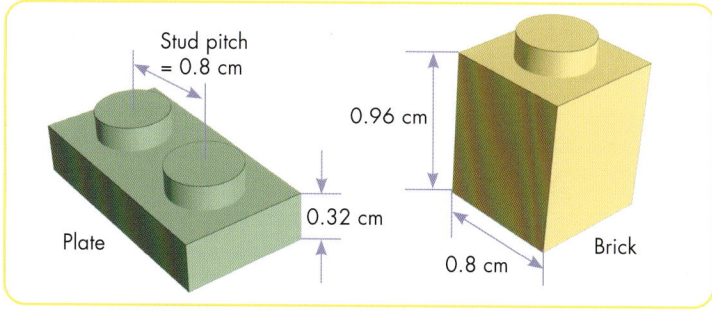

Figure 1-14: The dimensions of a 1×2 plate and a 1×1 brick

The 3:2.5 or 6:5 ratio of height to width of a LEGO brick, and the fact that a brick isn't a whole number of studs tall, can create a few challenges. We'll see these challenges come into play often in later chapters, especially when we start delving into sideways building.

DIGITAL BUILDING

Digital building has become increasingly popular among AFOLs. Instead of building a LEGO model using real pieces, you can choose to build a model virtually on your computer. There are several software options available for this purpose. The most popular option, and the one we'll focus on in this book, is BrickLink Studio, though there are also other programs such as LeoCAD and Mecabricks.

These programs don't just allow you to build a LEGO model digitally. The processing power available in modern computers also makes it possible to generate realistic renders of your models that are hard to distinguish from actual photographs. These programs can also automatically create the step-by-step instructions needed to build your model using real pieces, allowing you to share your designs with others.

ADVANTAGES

Why trade the tactile pleasures of working with real LEGO bricks for the comparatively sterile experience of putting a LEGO model together digitally? The answer comes down to cost and convenience. Most LEGO building software is available for free, so there's no upfront cost involved. By contrast, LEGO bricks themselves aren't exactly cheap, and not everyone has a large inventory of pieces available to turn any idea they may have into an actual model made of real bricks. Using software essentially allows you to have an unlimited inventory of LEGO pieces, albeit virtual ones, making it easier to create large, ambitious models.

The programs also provide shortcuts that can make the digital building process less tedious and repetitive than real-life building. For instance, if you're building a skyscraper with the same design used for each floor, you can simply create a submodel for a single floor and then copy and paste it as many times as needed.

DISADVANTAGES

One downside is that digital building tools can have a steep learning curve, especially if you don't have any experience using CAD software. Also, there's no guarantee that a digital LEGO model is even practical to build in real life. Basic connectivity and structural integrity checks are available in Studio, but they aren't foolproof. This is especially an issue for larger models, where you can get away with paying little attention to internal structural reinforcements when you're in the digital realm. If you try to turn the digital model into a real one, you may encounter unexpected challenges in ensuring the model is sturdy enough to be moved around.

Some people opt for a hybrid digital–physical approach. Digital building can be a great way to prototype a design and figure out what pieces are required. Then, once you're happy with the design, you can order those pieces through a website like BrickLink and build the model in real life.

THE LDRAW FORMAT

LDraw is a standard file format used to describe a digital LEGO model. An LDraw file is a text file with one line devoted to each of the parts that make up the model. This line has multiple fields containing information about the part type, its color, and its location and orientation in 3D space. The 3D coordinates are specified in terms of *LDUs*, or *LDraw units*. One LDU is roughly equivalent to 0.4 mm, which makes a 1×1 brick 20 LDUs wide by 24 LDUs tall. In other words, 1 stud is 20 LDUs, and 1 plate is 8 LDUs.

Most programs that allow you to digitally build LEGO models can export and import files in the LDraw format. The simplicity of the format also makes it possible to use programming languages such as Python, C++, and JavaScript to automate the creation of LEGO models such as mosaics and sculptures. We'll discuss these types of models in Chapters 8 and 9.

SUMMARY

This chapter surveyed the history and strengths of the LEGO system and outlined the basic types of elements available for building. You've learned some key terms, such as LEGO units of measure like the stud and the plate, that we'll repeatedly encounter as we explore all the possibilities that this wonderful system has to offer. Before we start getting into actual building techniques, however, it's useful to get a better understanding of the concepts of scale and proportion. That's where we'll turn next, in Chapter 2.

ALL ABOUT SCALE

The beauty of using LEGO as a creative medium is that it allows you to bring to life anything that your imagination can conjure up. Quite often, however, the subject of a LEGO build is something familiar, perhaps an ordinary house, an iconic landmark, or a scene from your favorite fictional universe. In this chapter, we'll discuss what's involved in designing a realistic LEGO replica of a recognizable structure. We'll focus on the concepts of scale and proportion—concepts integral to art and design that are applicable to LEGO builds as well.

SCALE AND PROPORTION

The *scale* of a LEGO model is its size relative to the real version it's trying to represent. For example, say you're creating a LEGO model of a building that's 10 feet tall, and you want your model to be 1 foot tall. You could express the model's scale as the ratio 1:10, where the first number refers to the size of your model and the second number refers to the size of the real building (see Figure 2-1).

10 units

1 unit

Figure 2-1: A 1:10 scale model

For the model to be accurate, this 1:10 ratio should apply to *all* the dimensions, not just the height. If the real building is 12 feet wide, your model would have to be 12/10 = 1.2 feet wide, or it won't look right; it will appear either too skinny or too squat compared to the real thing. In other words, the model won't be in *proportion*.

All the elements of a model need to have a consistent scale for the model to have the right proportions and be a faithful replica of the real thing. In the case of a building, it is not just the overall dimensions that need to have the same scale, but also smaller details like doors, windows, and shutters. If the length, width, and height of the building are all one-tenth the size of the original, but the door is only one-fifth the size, the door will be out of proportion with the rest of the model. Unless it's your intention to exaggerate the size of the door for aesthetic reasons, this discrepancy should be avoided.

It may be tempting to try to represent every little detail from the original version in your LEGO model for the sake of realism. However, this doesn't always add to the accuracy of your model, especially if you have no way of representing the smaller details at the right scale. For instance, if the door of a building has a knob with a diameter of 2.5 inches, it would have to be 0.25 inches in a 1:10 scale model. If the smallest diameter you can achieve for the knob in the scale model is 1 inch, it would be preferable to just omit the doorknob from the model rather than represent it with a size that will appear obviously incorrect.

> ## DIGITAL BUILDING TIP
>
> If you're building your model digitally in Studio, is there a way to figure out how big it will be in real life? There sure is! Just click the **Model** dropdown menu and select **Model Info**. This will open a window showing information about your model. From there, click the **Physical Information** tab to see the length, width, and height of your model (in studs, inches, and centimeters) as well as an estimate of its weight (in ounces and grams).

UPSCALED BRICKS

A good way to illustrate the concept of scale is to create an *upscaled brick*, a large replica of a LEGO brick built using regular-sized LEGO bricks. (This is one of the few cases where the LEGO model is *bigger* than the real-life object it's trying to represent; it's usually the other way around.) A common scale used for upscaled bricks is 3:1, meaning the overall size of the upscaled brick is three times that of a regular brick (in every dimension). Let's use a 2×4 brick as an example. It's 2 studs wide and 4 studs long, so a 3:1 upscaled version would be $2 \times 3 = 6$ studs wide and $4 \times 3 = 12$ studs long. Its height, not including the studs, should also be three times the height of a regular brick.

To build the upscaled brick, we could use regular bricks (such as 1×4 and 1×6 bricks) for the outer walls. The scale calls for the model to be 3 bricks high, so let's start there (see Figure 2-2).

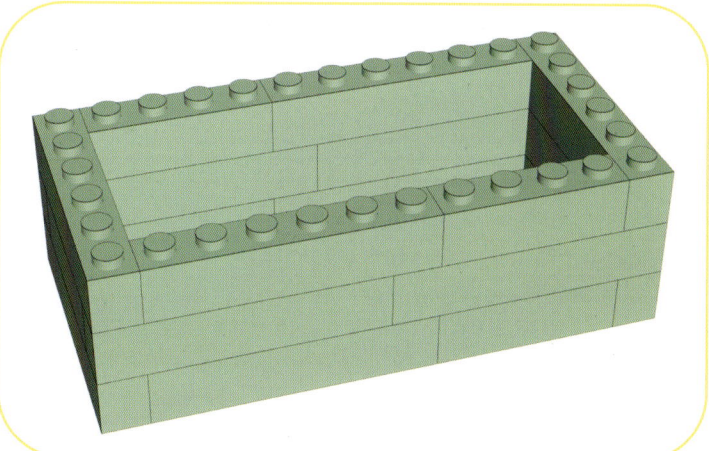

Figure 2-2: The walls of the upscaled brick

The next step might be to use plates and tiles to close off the hole in the middle of the upscaled brick and create a flat surface across the top. There's a problem, however: to honor our 3:1 scale, the model should be exactly 3 bricks tall (not counting the studs on top). If we start adding plates or tiles on top of the structure in Figure 2-2, the model will be too tall, with the height out of proportion relative to the length and width.

Remember that one layer of bricks has the same height as three layers of plates or tiles. We can therefore replace the top layer of bricks with two layers of plates (to close the hole), plus one layer of tiles (to cover up the exposed studs). There are many sizes of plates we could choose from, but to keep things simple, let's use large 6×12 plates that span the whole size of the model. We need only two, one for each of the two plate layers (see Figure 2-3).

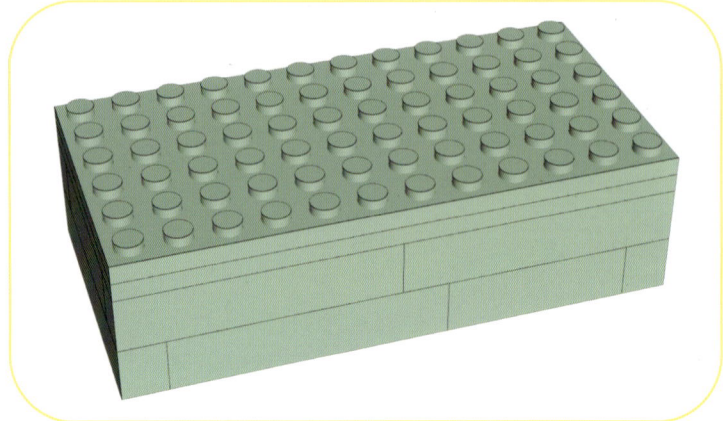

Figure 2-3: Replacing the top layer of bricks with two 6×12 plates

Once we add a layer of tiles on top of the plates, the height will be exactly right. But let's pause here for a moment and think about how we'll represent the bumps on top of our upscaled 2×4 brick. On a regular-sized LEGO brick, the bump is cylindrical with a diameter of 0.48 cm (1.5 plates) and a height of 0.16 cm (0.5 plates), as shown in Figure 2-4.

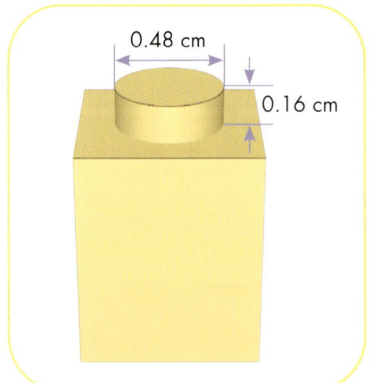

Figure 2-4: The dimensions of the stud (bump) on a 1×1 brick

At our 3:1 scale, the bumps should theoretically be 1.5 × 3 = 4.5 plates wide and 0.5 × 3 = 1.5 plates high. This is a problem: there are no LEGO elements that match these dimensions. The closest we can get, while also matching the round sides and flat top of a bump, is to stack a 2×2 round tile (#14769) on top of a 2×2 round plate (#4032). This combination has a diameter of 5 plates (2 studs) and a height of 2 plates. We'll be off by half a plate in each direction, but given that the overall model is 6×12 studs (or 15×30 plates), hopefully this won't make much of a difference.

Now let's figure out how to attach the upscaled bumps to the top of the upscaled brick. As we discussed in Chapter 1, a 2×4 brick is equivalent to eight 1×1 bricks placed in two rows of four. Both occupy 2 × 4 = 8 units in the regular square grid that forms the basis of the LEGO system, with the grid unit size being equal to 1 stud. Each bump on the top surface of a 2×4 brick is centered within a grid unit. In our upscaled version, this grid unit has a size of 3×3 studs. Imagine dividing the 6×12 surface on top of our upscaled brick into eight squares that are each 3×3 studs. The upscaled bumps must be attached centrally within each 3×3 square, so we need to leave an exposed stud (in the form of a 1×1 plate) in the middle of each square, as in Figure 2-5.

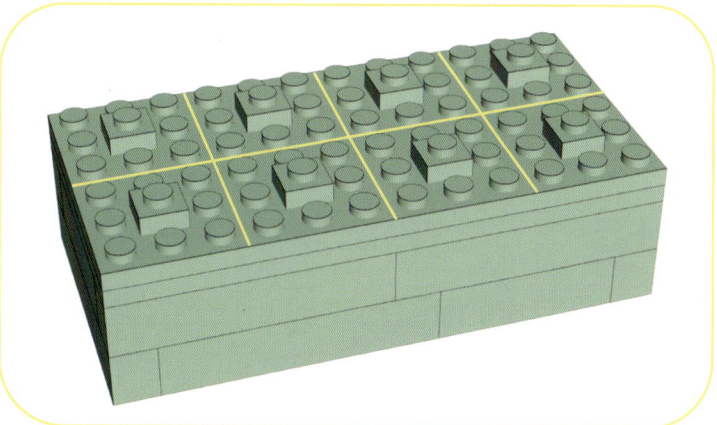

Figure 2-5: The eight 3×3 squares on top of our upscaled brick
are each equivalent to a stud on a regular brick.

Now we can fill in the rest of the top layer of the brick with tiles so that its surface appears smooth (Figure 2-6). Between the plates and the tiles, the brick now has the correct height, along with attachment points for the upscaled bumps.

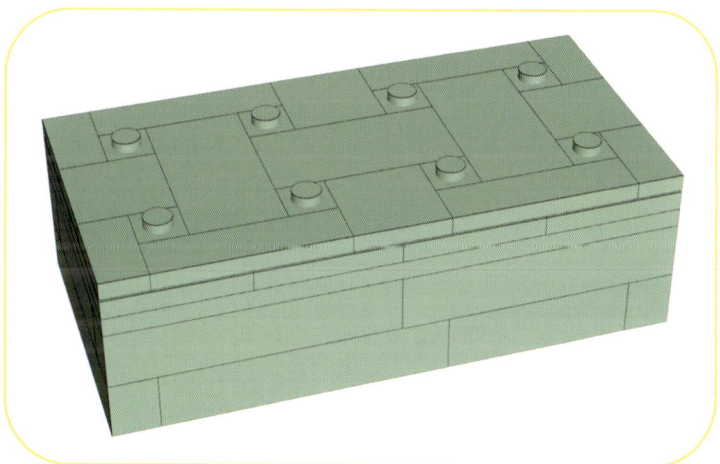

Figure 2-6: Covering the top surface of the upscaled brick with tiles

A nice thing about the 2×2 round plate we're using for the bottom layer of each bump is that it has a Technic axle hole in the middle, the underside of which doubles as an anti-stud (as shown in Figure 2-7). We can connect that anti-stud to the exposed stud in the middle of each 3×3 square to perfectly center each bump.

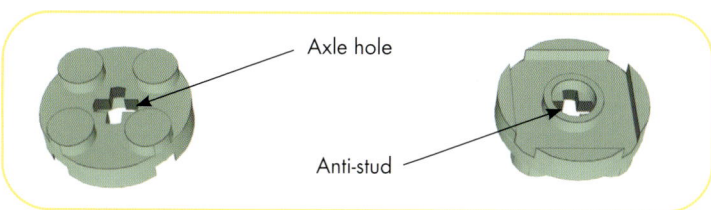

Figure 2-7: The top and underside of a 2×2 round plate

Attach all the bumps, and voilà: we have a 3:1 upscaled version that does indeed look very much like a regular 2×4 brick (Figure 2-8).

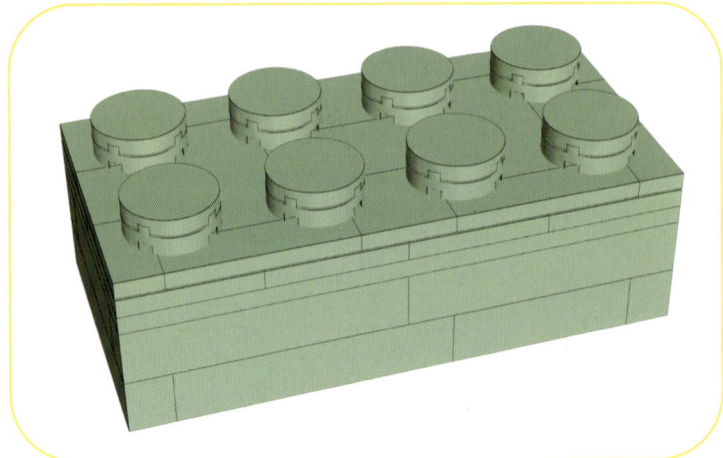

Figure 2-8: An upscaled 2×4 brick

As this example has shown, there's often some rounding involved in scale modeling. In this case, we had to fudge the dimensions of the bumps a little bit. As long as the proportions are close, these sorts of compromises are perfectly acceptable and shouldn't detract from the creation of a reasonably faithful replica of the original.

MINIFIG SCALE

When we categorize the scale of a painting or sculpture, we humans typically use ourselves and the world around us for reference. A *life-size* sculpture has a 1:1 scale, meaning it's the same size as the real-life person or object it's depicting (picture a figure in a wax museum, for example). Terms like *large scale* (used for Michelangelo's *David*), *monumental scale* (used for Mount Rushmore), *small scale*, and *miniature scale* are all relative to the life-size scale. In the LEGO world, building at a (human) life-size scale is often impractical, but that's okay, because the LEGO world has its own versions of humans that models can be scaled to: minifigures.

Introduced in 1978, *minifigures* (*minifigs* for short) are small figurines that are either packaged as a part of LEGO sets or sold separately. They're composed of body parts (head, torso, arms, hands, hips, and legs) that are interchangeable (Figure 2-9). These parts come in different colors and with various designs printed on them, making minifigures highly customizable. There's also a wealth of accessories available that can be added to minifigures to customize them even more. These include hair and headgear (that can be attached to the stud atop a minifig's head) and various utensils, weapons, and other objects that can be attached to minifig hands (which are essentially clips). Minifigs have anti-studs under their feet as well as on the back of their legs, allowing them to be attached to LEGO studs in either a standing or sitting position.

If minifigs are the LEGO equivalent of humans, it makes sense to build models where real-world structures are scaled down to their size—in other words, *minifig scale*. Unfortunately, it isn't always easy to come up with a single ratio for minifig scale. The problem lies in the proportions of the minifigs themselves.

22

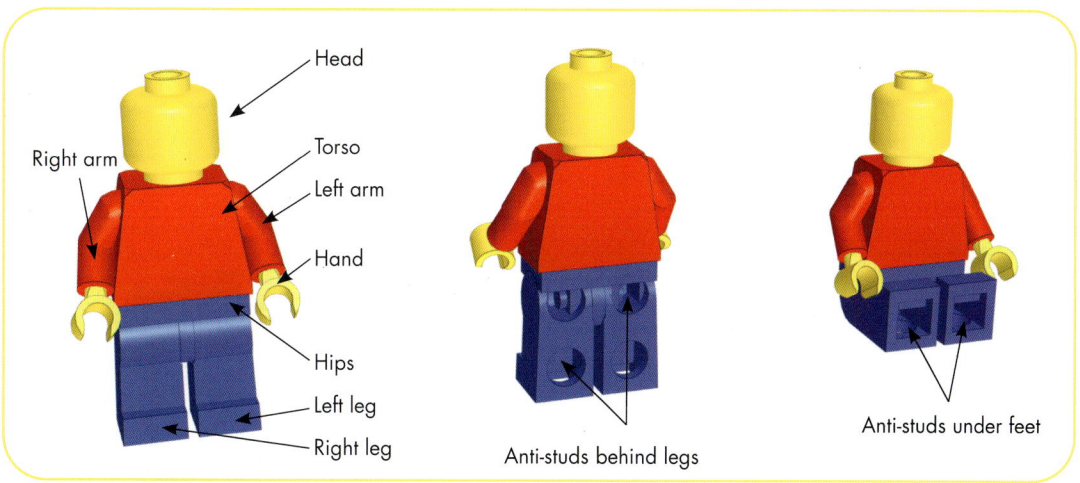

Figure 2-9: The parts of a LEGO minifig

A standard minifigure is 4 bricks tall (3.84 cm), not counting the stud on top of its head. It's exactly 4 cm tall if you include the stud. A minifig's torso is 2 studs wide (1.6 cm), not including the arms on either side (see Figure 2-10). An average human is about 170 cm tall, so based on just height, minifig scale translates to about 1:42. However, the average width of a human torso measured at the shoulders is 43 cm. Based on that measurement, minifig scale would be closer to 1:27.

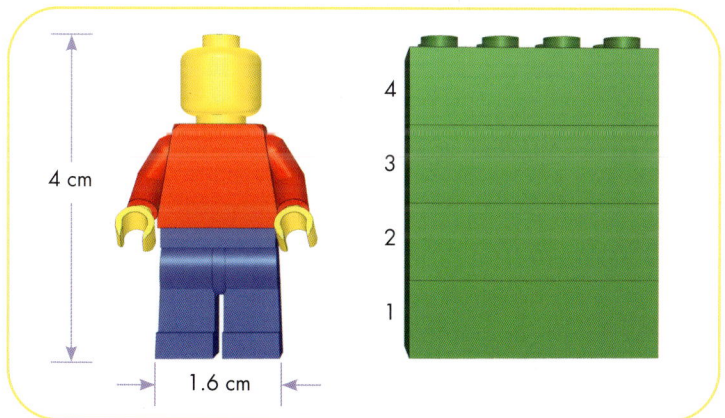

Figure 2-10: The dimensions of a standard minifigure

Clearly, a minifigure is a rather cartoonish representation of a human being, with overall proportions that are squatter than the average person. If you want to scale something down to minifig size, ratios in the range of 1:25 to 1:50 should work. That said, most LEGO builders think of something closer to a height ratio of 1:42 when they talk about minifig scale.

OTHER SCALES

Other scales used for LEGO builds are generally categorized using minifig scale as a reference. For example, any scale smaller than minifig scale is called *microscale*. This term encompasses the entire gamut of scales smaller than around 1:50. It's likely that the majority of LEGO versions of famous landmarks that you've seen (both official sets and MOCs) can be classified as microscale models.

23

Another common scale is *Miniland scale*, which is used for most of the large models you can find in the Miniland displays at various LEGOLAND Parks. At this scale, the figures representing humans are built using actual bricks and are typically about 11 bricks (10.56 cm) tall. Figure 2-11 shows an example of a Miniland figure.

A height of 11 bricks puts Miniland scale at around 1:16, but building human figures using regular bricks offers many more opportunities for customization, including in the height. So slightly smaller scales, like 1:20, also qualify as Miniland scale.

HOW TO CHOOSE A SCALE

Figure 2-11: A Miniland-scale figure

There's a trade-off associated with scale. The bigger the scale, the more accurately you can represent all the elements of the original version in your LEGO model, but too large a scale can result in a massive, unwieldy model with a prohibitively high part count and cost. On the flip side, using too small a scale might force you to compromise on accuracy more than you find acceptable. The trick, as with everything, is finding the sweet spot that works best for you.

To illustrate how you might approach deciding on a scale, let's look at two examples. First, we'll try to develop a LEGO model of a simple two-story house, the sort you might find in a typical North American suburb. Then we'll set our sights on one of the most famous buildings in the world: the Empire State Building. In both cases, we'll see how the dimensions of a real building can be translated into LEGO dimensions at the desired scale.

MODELING A HOUSE

Let's say we want to model the two-story house shown in Figure 2-12. It has a footprint of 40×30 feet and a height of 20 feet up to the front and back roofline (making each floor 10 feet tall). These are pretty typical dimensions for a house, but they're just an example. If you have a specific house (or any other building) in mind that you'd like to model, you can try taking digital measurements of its dimensions while viewing it in an online tool like Google Earth. (We'll discuss how to do this in more detail later in the chapter.) Or, in the case of your own home, perhaps you have access to a floor plan drawing.

There are many possible scales we could use for this model. One natural choice is to build a version of the house that our LEGO counterparts (namely, minifigs) can live in. In this case, the right scale to use would be minifig scale, which, as we've discussed, is around 1:42. With this scale in mind, let's walk through the process of developing the model.

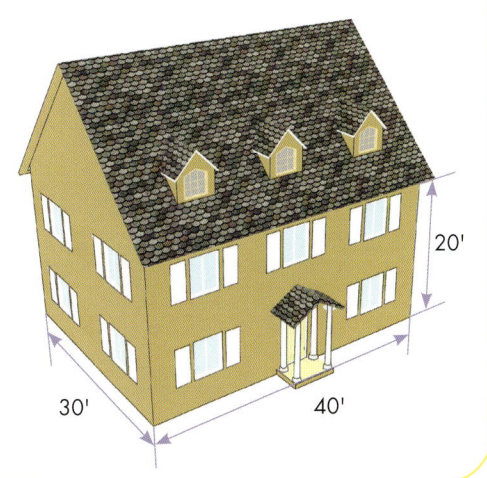

Figure 2-12: A simple house

DETERMINING THE FOOTPRINT

The first thing we should do after choosing the model's scale is figure out the size of the model's footprint in studs. For that, we need three steps:

1. Convert the dimensions to centimeters. In this case, 40×30 feet is equivalent to about 1,219×914 cm.
2. Divide by the scale factor to get the LEGO model's dimensions in centimeters. In this case, we need to divide each dimension by 42. That gives us a footprint of around 29×21.75 cm.
3. Divide by 0.8 cm (the equivalent of 1 stud) to get the dimensions in studs. This works out to around 36×27 studs.

But there's a catch: when we have a sloping roof, an even number of studs always works better for the dimension along the slope. This is based on the roof pieces we'll need to use for the apex of the roof, where the two sloping sides meet. (We'll cover building the roof in Chapter 3.) So we'll round the depth of the model from 27 up to 28 studs. As with the bumps on top of the upscaled brick, this is another case where a bit of fudging is necessary, but 1 stud shouldn't make much of a difference.

PLACING THE DOOR AND WINDOWS

Next, let's think about the placement and dimensions of the door and windows. On the first floor, there are two windows and a door on the front wall, and two windows on each side wall. If we use the standard door and window pieces that are available in the LEGO catalog, one limitation we'll run into is the fact that these pieces are available only in even stud widths. But looking at the sizes of the door and windows in the house relative to the wall segments in between them, it appears that a width of 4 studs will work for each of the door and window openings. (We could confirm this using actual measurements if needed.)

For the front door of the house, we'll use a 1×4×6 door frame (#60596) fitted with a 1×4×6 door piece with a stud handle (#60616). We'll use four 1×2×2 window frame pieces (#60592) joined together in two rows of two for each of the windows. These window frames will be fitted with glass window inserts (#60601) in a transparent brown color.

With 12 studs out of 36 taken up by the windows and door, that leaves 24 studs on the front of the house for solid wall segments. Since the house is laid out symmetrically, we can split those 24 studs into four segments that are 5, 7, 7, and 5 studs wide. On the sides of the house, if the windows are again 4 studs wide, we can make the wall segments 5, 10, and 5 studs, as shown in Figure 2-13.

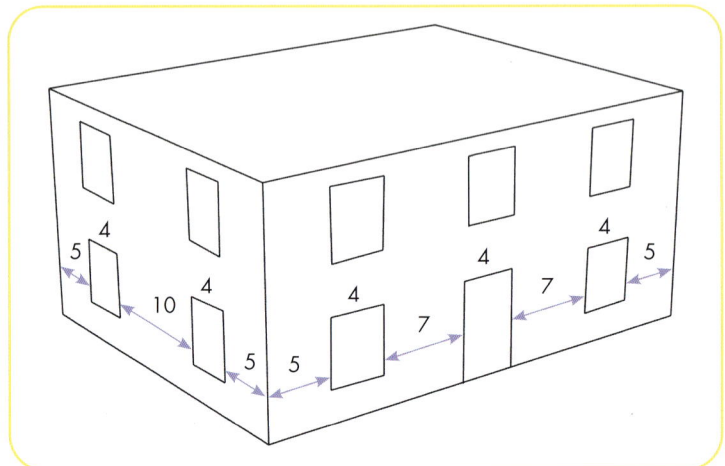

Figure 2-13: The dimensions of the wall segments, windows, and door for the model house

CALCULATING THE HEIGHT

How many layers of bricks do we need for each floor of the house? To figure that out, take the height of each floor (10 feet, or 305 cm) and divide it by the scale factor 42. We get around 7 cm. The height of each brick is 0.96 cm, which is close enough to 1 cm. Therefore, each floor of the house needs to be 7 bricks tall. The door piece is 6 bricks tall, while the four-panel windows are 4 bricks tall. We'll want to put the door at ground level, with a single layer of bricks above it to reach the full 7-brick height. For the windows, we'll want two layers of bricks below and one above.

BUILDING THE FRAME

Now we have all the information we need to lay out the pieces and create the basic frame of the house. Here are the steps to take:

1. Start building the first floor by placing bricks for the first two layers, leaving an opening in the middle of the front wall for the door (Figure 2-14).
2. Attach the door and window pieces (Figure 2-15). Remember the widths we discussed for the wall segments in between: 5, 7, 7, 5 for the front and back, and 5, 10, 5 for the sides.
3. Build up the wall sections that fill the gaps between the windows and door (Figure 2-16).
4. Add a final layer of bricks around the perimeter of the house (Figure 2-17). This layer locks the windows and door into place.
5. Create the second floor. It's identical to the first floor, except the door should be replaced with another window (Figure 2-18).
6. Stack the two floors on top of each other (Figure 2-19).

26

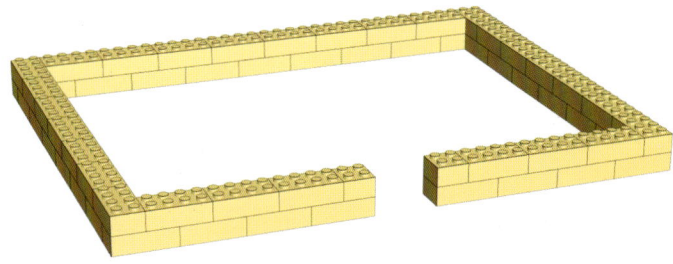

Figure 2-14: Building the first two layers

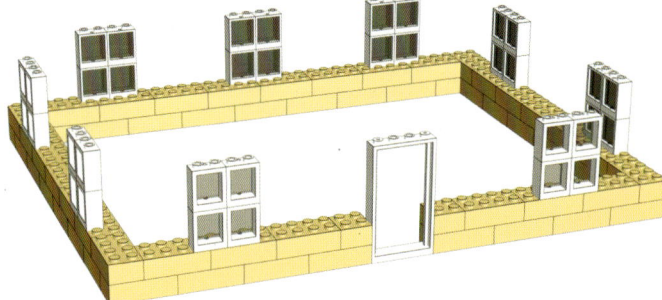

Figure 2-15: Attaching the window and door pieces

Figure 2-16: Adding the wall segments between the windows and door

27

Figure 2-17: Adding the final layer for the first floor

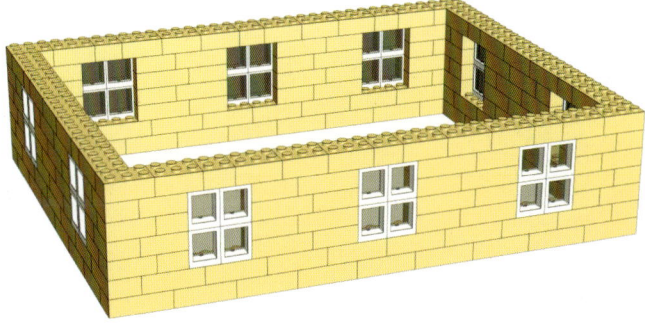

Figure 2-18: Building the second floor

Figure 2-19: Both floors stacked

We'll pause the model here and look at adding the roof and other cosmetic details like the window shutters as we explore the relevant building techniques in the chapters that follow.

MODELING A SKYSCRAPER

Now that we've seen how the scale calculations work for a generic house, let's think a bit bigger: What would it take to build a LEGO model of the Empire State Building? Completed in 1931, this skyscraper (shown in Figure 2-20) was the tallest building in the world for about 40 years. It's since been surpassed in height, but it remains one of the world's most iconic landmarks. With such a recognizable structure, accuracy will surely be important. If the proportions are off, people will probably notice.

Figure 2-20: The Empire State Building

The shape of the Empire State Building isn't as simple as that of the house. It has a wide base topped by a tower that tapers as it rises. This distinctive tapered shape, which came to be associated with the Art Deco–style skyscrapers built during the early 1930s, wasn't just an aesthetic choice; it was necessitated by the zoning restrictions in place in New York during that time, which were intended to prevent the new skyscrapers rising in the city from blocking too much sunlight on the streets below.

BREAKING DOWN THE SHAPE

The best way to approach designing a LEGO model of something with a complex shape such as the Empire State Building is to break that shape into simpler components, as shown in Figure 2-21.

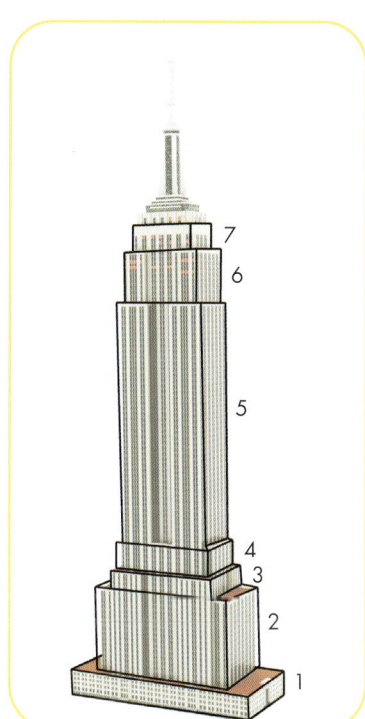

Figure 2-21: Breaking the structure's complex shape into simpler components

We can then use the biggest or most dominant component for our scale calculations.

Loosely speaking, if we don't include the spire at the top, we can think of the shape of the Empire State Building as a stack of seven rectangular prisms, including the wide base, that get progressively smaller as the building rises. These prisms should roughly correspond to the different sections making up our LEGO model.

Luckily, the LEGO medium is well suited to creating blocky shapes like rectangular prisms, so the Empire State Building lends itself quite naturally to being represented in LEGO. It's no wonder, then, that there are numerous LEGO models of the Empire State Building, including two official LEGO sets (21002 and 21046) and many MOCs. These models vary wildly in scale, from tiny (under 8 inches tall) to massive (over 25 feet tall).

DETERMINING THE SCALE

With such a wide range of options, what scale should we choose for our version of the Empire State Building? Of course, there's no single right answer, but let's think through some possibilities. First, say we wanted to build at minifig scale. How big would the model have to be? The real Empire State Building is 1,454 feet tall. Divide that by 42 and we get about 35. In other words, a proper minifig-scale version of the Empire State Building would have to be 35 feet tall! That's even bigger than those massive models you see in a Miniland. Unless you have a very tall ladder and an infinite supply of bricks (and time), this clearly isn't a practical option.

Perhaps we need a different approach. Instead of picking an arbitrary scale factor and calculating the resulting height of the model, let's focus on some overall goals for the model and work backward from there to identify the optimal scale. In general, it would be nice to build a model that's reasonably accurate, but accuracy can be very subjective. We'll have to decide what features of the original version we want to represent accurately and figure out how that squares with our constraints in terms of the overall size, part count, and cost of the model.

CAPTURING THE FACADE

Many people would be content with representing just the outer shape of the Empire State Building accurately—the way it tapers from bottom to top. Let's set a goal to go one step further than that and try to accurately represent the distinctive window arrangement on the Empire State Building's facade. Take a closer look at the tallest section of the building (section 5 from Figure 2-21), shown in Figure 2-22.

Figure 2-22: A closer look at the building's window arrangement

Notice that the front face of the building features columns of windows, organized into groups of two or three, with wall segments in between. Looking from left to right, there's a wall segment, then two columns of windows, then another wall segment, then three columns of windows, and so on. If we represent each window with an x and each wall segment with a dash, the whole arrangement looks like this:

-xx-xxx-xx-**xx-xx-xx**-xx-xxx-xx-

The middle portion of the facade (shown in bold) is recessed, or set back from the main face of the building.

The nice thing about this building is that all the individual windows appear to have the same width, which is also similar to the width of the wall segments. The smallest possible model that could represent this arrangement accurately would therefore use 1 stud for each window or wall segment. Let's go with that, on the assumption that we don't want to have to make the model any larger (and thus costlier) than necessary. This would add up to a total width of 30 studs on the longer sides. Meanwhile, the shorter sides of the building have seven groups of two windows, with a wall segment between each group, like this:

-xx-xx-xx-xx-xx-xx-xx-

That works out to a total of 22 studs for each short side.

TAKING REAL MEASUREMENTS

We've determined that the tallest section of the model will have a footprint of 30×22 studs, but what scale does that represent? To figure that out, we need measurements from the real

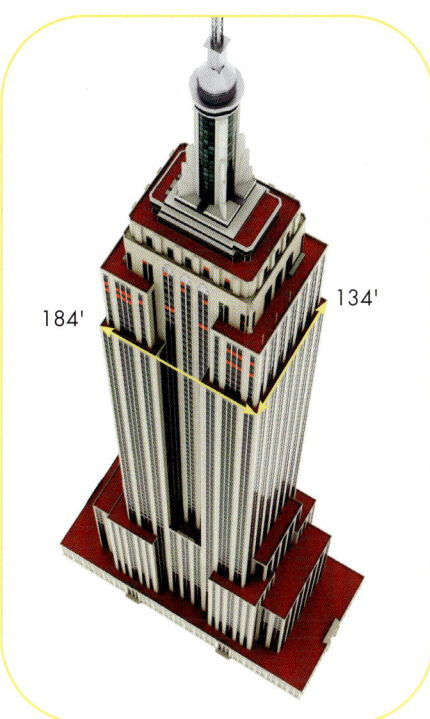

Figure 2-23: Measuring the footprint of the tallest section

building. For something like the Empire State Building, you may be able to find architectural drawings online, but a simpler option is to use Google Earth. Zoom in on the 2D aerial view of the section of the building you're interested in, and use the Measure tool (the ruler icon) to find the distance from one point to another. Using this method, we can determine that the footprint of the tallest section of the Empire State Building is 184×134 feet (see Figure 2-23).

And there we have it. Our model will use 30 studs to represent 184 feet. Each stud is 0.8 cm wide, so 30 studs would be 30 × 0.8 = 24 cm wide. We need to convert the width of the real building into metric units as well: 184 feet is roughly 5,608 cm. Therefore, our scale ends up being 24:5608, or roughly 1:230. We arrive at the same number using the dimensions of the shorter side: 22 studs = 17.6 cm, and 134 feet = 4,084 cm. The scale is 17.6:4084, which again is about 1:230.

Once we have the scale, it's easy to figure out how tall the model should be: just divide the total height of the Empire State Building (1,454 feet = 17,448 inches) by 230 and you get 76 inches (6 feet, 4 inches). Considering we started at 35 feet, this is starting to sound more achievable.

SCALING EACH FLOOR

Next, we need to figure out how tall each floor of the building should be in terms of brick heights. We can once again use Google Earth and get the difference in elevation between the top of the section of interest (917 feet) and its bottom (415 feet, as shown in Figure 2-24). That's a height of 502 feet.

Some online research indicates this section of the building comprises 42 floors, from floor 30 through floor 71. The height of each floor is then 502/42 = 12 feet, or 366 cm. Applying our 1:230 scale factor, each floor of our model should be 366/230 = 1.59 cm tall. This is equivalent to 1.59/0.96 = 1.66 bricks.

It might be tempting to round 1.66 up to 2 and use two layers of bricks for each floor in the model.

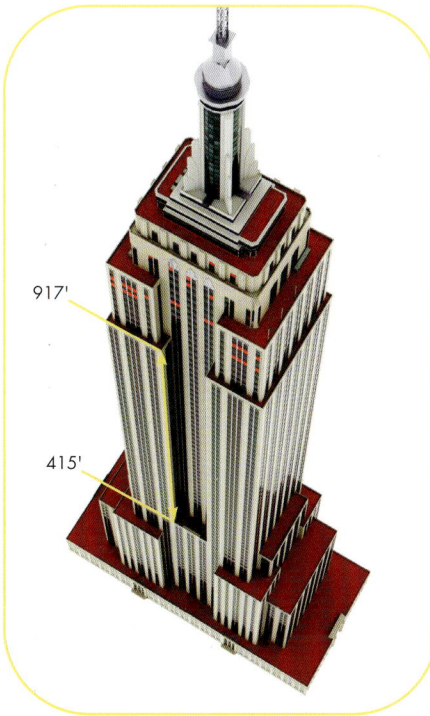

Figure 2-24: Measuring the height of the biggest section

This sort of rounding would work just fine for a model of a house that has one or two floors. But given that the Empire State Building has 86 floors (excluding the spire), it's important to understand how the rounding affects the model's proportions. Using 2 bricks per floor instead of 1.66 would make our model around 92 inches tall, instead of the 76 inches we originally planned, since (76 × 2)/1.66 = 92. The effect of the rounding would add up, making our model appear stretched compared to the real building.

We have two options to address this. First, since plates are a third as tall as bricks, we can avoid rounding altogether by making each floor 5 plates (or 1 brick and 2 plates) tall. That works out to exactly the 1.66-brick height we need. This will drive up the part count in the model and increase its complexity, however. Making a layer of bricks plus two layers of plates is a lot more involved than simply making two layers of bricks.

Another option is to use 2 bricks per floor (which will simplify the construction of the model) but adjust the number of floors by a factor of 1.66/2 = 5/6 to keep the overall height of the building the same. In other words, instead of having 42 floors that are each 5 plates high in the tallest section of the model, we'll have 42 × 5/6 = 35 floors that are each 2 bricks high. The number of floors would have to be scaled down in a similar way in the remaining sections of the model. Compromises like this may not be ideal, but in this case, the significant reduction in the part count and the simplification of the building process make it worthwhile.

Full disclosure: what I've outlined here is exactly the thought process I went through to determine the scale of my own model of the Empire State Building. We won't get into the process of actually building it right now, but we'll revisit the model in later chapters as we discuss some of the concepts and techniques required to realize it.

THE TRADE-OFFS OF SCALE

The 1:230 scale represents a sort of sweet spot for an Empire State Building model in terms of the priorities we laid out: it's large enough to accurately capture some of the details of the facade, but not so large as to be completely absurd. That said, not everyone has the wherewithal or the interest to build a LEGO model that's over 6 feet tall. A smaller model would have the advantage of requiring fewer pieces and being easier (and less expensive) to build. But it would also require trade-offs in terms of the level of detail.

DIFFERENT EMPIRE STATE BUILDINGS

To illustrate the trade-offs at play in choosing a scale, let's consider how LEGO's two official Empire State Building sets compare to our model. The original LEGO Architecture set (21002) of the Empire State Building uses just 77 pieces and stands a mere 7.4 inches tall (see Figure 2-25). That would put the scale of the model at around 1:2400 (less than a

Figure 2-25: The LEGO Empire State Building set 21002

tenth the scale of our 1:230 model). At this tiny scale, the only thing we can hope to represent (albeit not very accurately) is the outer shape of the building. This is because the 1:2400 scale

Figure 2-26: The LEGO Empire State Building set 21046

simply doesn't allow for any other details (like windows) to be represented.

The newer Architecture set (21046) of the Empire State Building, released in 2019, takes things to the next level (see Figure 2-26). It stands 21 inches tall, making the scale around 1:800. Here, the outer shape of the building can be represented a little more accurately, but once again, there's no way to correctly represent the window configurations—the main section is just 8 studs wide. Instead, the designer of the set was able to create the appearance of more detail by lining almost the entire facade of the model with 1×2 grille tiles (#2412). The use of 1×2 grille tiles to mimic window detail is an effective technique that was popularized by Spencer Rezkalla and Rocco Buttliere with their 1:650 renditions of well-known skyscrapers. The downside is that this significantly increases the complexity of the design. Instead of being able to simply stack bricks and plates, covering the facade with grille tiles requires sideways building (SNOT) techniques. (We'll cover some of these techniques in Chapter 5.)

Figure 2-27 shows both official LEGO versions of the Empire State Building next to a 1:230 scale model to give you a sense of the relative sizes of these models.

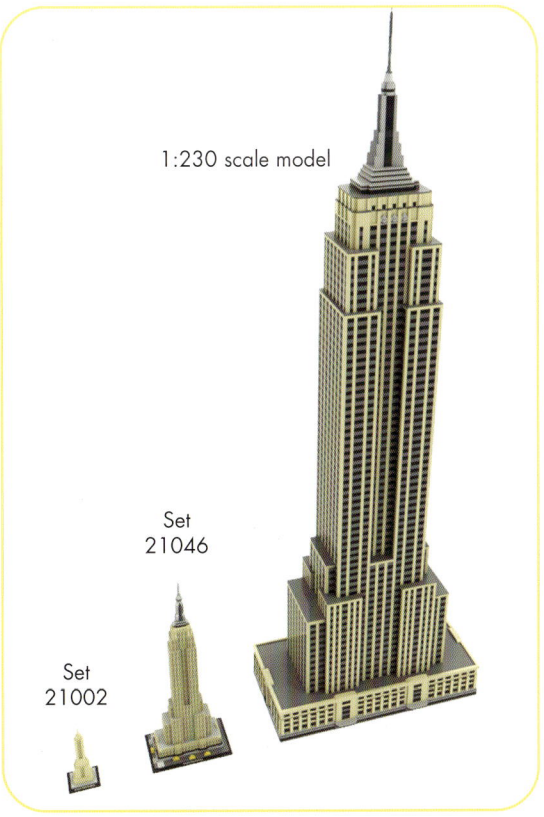

1:230 scale model

Set 21046

Set 21002

Figure 2-27: Models of the Empire State Building built to three different scales

DIFFERENT TAJ MAHALS

For another example of the trade-offs involved in choosing a scale, consider LEGO's two official models of the Taj Mahal. You wouldn't be able to pull off a convincing model of this iconic structure without finding a way to represent its massive central dome. Let's see how LEGO approached this challenge in its two models, which are shown in Figure 2-28.

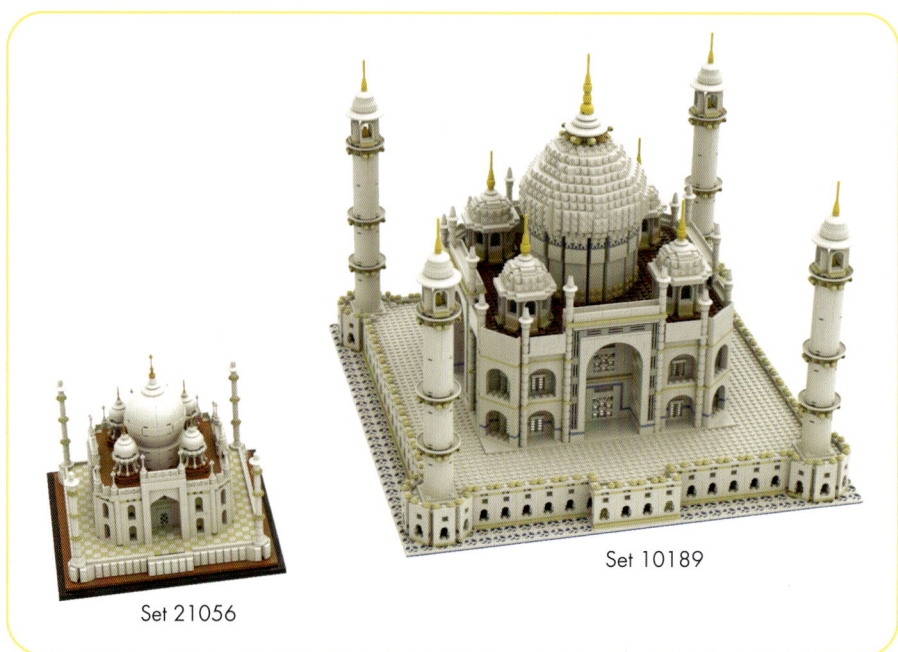

Set 10189

Set 21056

Figure 2-28: Two official LEGO sets of the Taj Mahal

The LEGO Architecture set of the Taj Mahal (21056) was released in June 2021 and uses the smaller scale of the two models. Here, the designer used four 4×4 round corner bricks (#49612) to form a dome with a diameter of 8 studs. When you use specialized pieces like this in your model for a shape that can't otherwise be created easily, those pieces end up dictating the rest of your decisions. In the case of this Taj Mahal set, it's likely that the size of the dome determined the scale for the entire model, which happens to be around 1:400.

The Creator Expert set of the Taj Mahal (10189), released in 2008, has a bigger scale (around 1:200), with the model occupying a footprint of 64×64 studs. Using round corner bricks isn't an option here, given the larger size of the dome, so the dome instead has to be built using regular plates. The technique for doing so will be covered in Chapter 7, but you can see how the bigger scale allows for more of the details from the actual Taj Mahal to be represented in the LEGO model.

As both the Empire State Building and Taj Mahal examples have shown, using a smaller scale in a LEGO model may translate to a smaller part count, but it doesn't always simplify the design of the model. In fact, every scale that you pick for your model can come with unique constraints and challenges that you'll need to overcome creatively.

SUMMARY

This chapter has illustrated how the concepts of scale and proportion come into play for LEGO builds. We've looked at the process of determining the right scale for a LEGO model and discussed some of the trade-offs involved in making that decision. Once you've decided on a scale, the next step is to figure out how to actually build the model. We'll begin delving into that in Chapter 3, which covers basic building techniques and sets the stage for the more complex techniques in the chapters that follow.

BASIC LEGO TECHNIQUES

For thousands of years, people have used the art of masonry to build all kinds of magnificent real-world structures out of bricks. While the bricks themselves may have changed over the years, the techniques for arranging them, known as *bond patterns*, have not. As another brick-based art form, LEGO building has a lot to learn from masonry. In this chapter, we'll discuss a few basic masonry-inspired techniques that encapsulate the best practices for laying out the bricks (or plates) in a LEGO model. Whether you're building a simple wall or a complex sculpture, these techniques can go a long way toward ensuring that your model holds together well and is as sturdy as possible.

OVERLAPPING BRICKS FOR STRENGTH

The simplest masonry bond pattern is the *stacked bond*, created by stacking bricks directly on top of each other, with the vertical joints between bricks all perfectly aligned (see Figure 3-1). However, while this pattern may be aesthetically pleasing, it's weak from a structural point of view. Each vertical stack of bricks stands as its own unit and isn't tied to the other stacks.

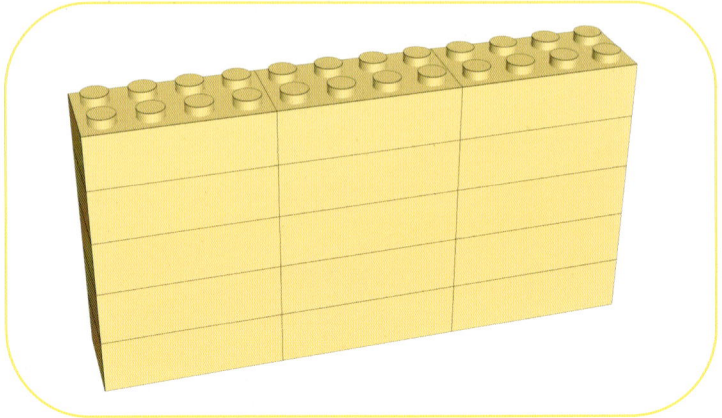

Figure 3-1: The stacked bond pattern

THE STAGGERED BOND

A stronger alternative is the *running*, or *staggered*, bond pattern (see Figure 3-2). Here, each row of bricks is offset by half a brick length relative to the row above and below it. The resulting overlap between the bricks in the different rows interlocks all the bricks together into a single structural unit, yielding a much sturdier wall.

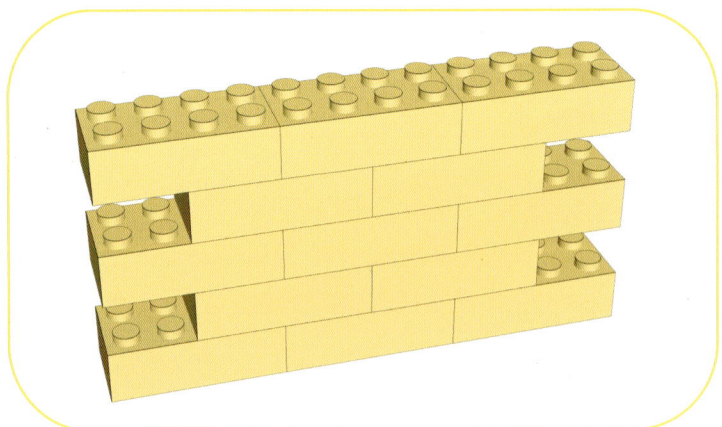

Figure 3-2: The staggered bond pattern

If you have some 2×4 bricks lying around, try creating a wall by simply stacking them with no overlap, and test with a flick of your finger how sturdy (or not sturdy) it is (see Figure 3-3). You'll find that this wall is not nearly as stable as one built as a fully interlocked structure using the staggered bond pattern, with the bricks in every other layer offset by 2 studs.

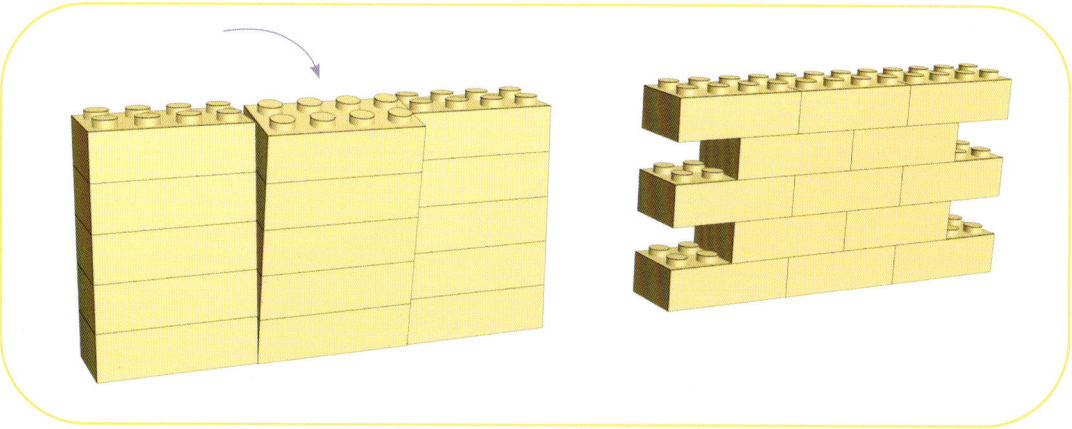

Figure 3-3: A stacked versus staggered bond

DIGITAL BUILDING TIP

The "finger flick" test obviously isn't possible when you're building a model digitally using Studio. Instead, there's a *connectivity check* that identifies parts that aren't fully connected to the rest of the model. To run this check, click the **Stability** button in the toolbar and switch to the **Connectivity** tab in the resulting box. This will highlight any detached sections in the model. A staggered bond pattern will pass this check, while a stacked bond pattern will be flagged for having multiple detached sections. Connectivity checks are a great way to ensure that your digital model will be sturdy in real life.

L-JUNCTIONS AND T-JUNCTIONS

The staggered bond pattern lends itself naturally to 90-degree corners in walls, also known as *L-junctions*, as shown in Figure 3-4. The staggered pattern can simply continue around the corner.

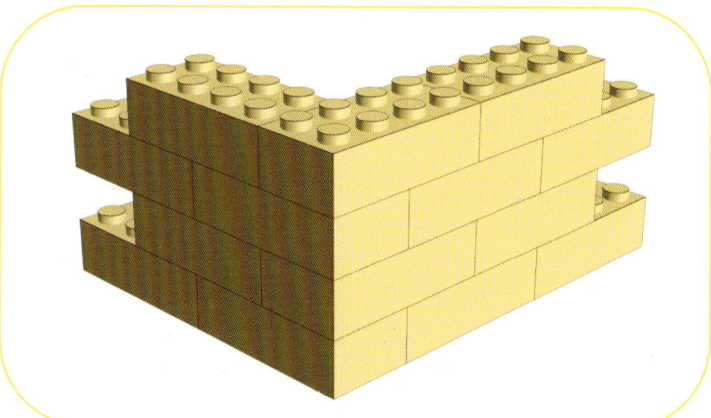

Figure 3-4: An L-junction with a staggered bond

However, the pattern needs to be adjusted a bit for *T-junctions*, places where an interior wall meets the middle of an exterior wall. To be able to interlock the perpendicular interior wall segment with the main exterior wall without disrupting the overall pattern, we need to mix in some shorter bricks (such as the 2×3 bricks used in Figure 3-5) on every other layer at the junction.

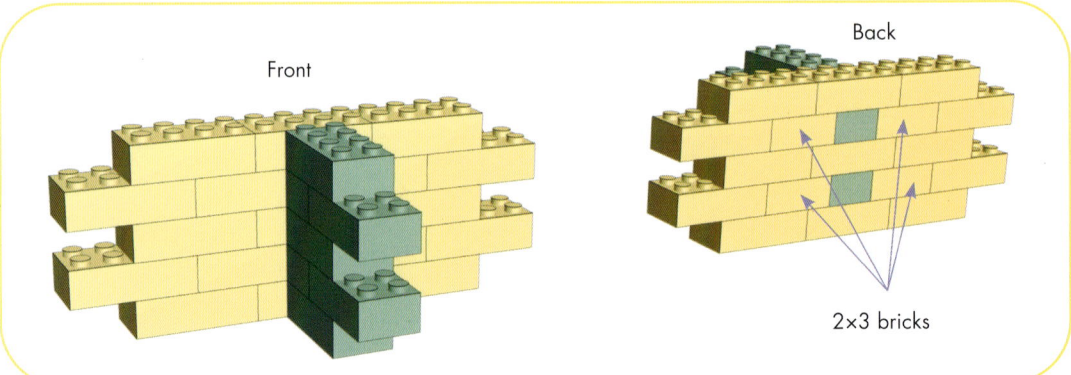

Figure 3-5: A T-junction with a staggered bond

If you take a closer look at the walls of the house we started building in Chapter 2, you'll see that we mostly used 2×4 bricks in a staggered bond pattern in order to avoid vertically aligned joints in consecutive layers wherever possible. However, we can't have a continuous 2×4 staggered bond pattern because the walls are interrupted by the windows and door. This forces us to use some shorter 2×2 and 2×3 bricks, and to offset some of the vertical joints by 1 stud rather than 2 (see Figure 3-6).

Figure 3-6: A modified staggered bond pattern to accommodate window and door placement. The top layer of bricks locks the windows and door in place.

Vertically aligned joints are unavoidable on either side of the window and door openings. This makes the layer of bricks immediately above these openings especially important: the joints in this layer must be offset from the edges of the openings to lock the windows and door in place. Above the door, for example, we use two 2×4 bricks that meet in the middle of the doorframe. Meanwhile, we center a 2×2 brick above each window, flanked by 2×3 bricks.

CLUTCH POWER AS MORTAR

Masonry walls usually use mortar to bond the bricks together. The mortar goes both in the horizontal joints between successive layers of bricks and in the vertical joints between adjacent bricks in the same layer. In the case of LEGO models, there's no bond whatsoever between bricks placed next to each other. The only thing holding a LEGO model together—the "mortar" of the LEGO world—is the clutch power between each layer and the layers immediately above and below it, as shown in the following figure.

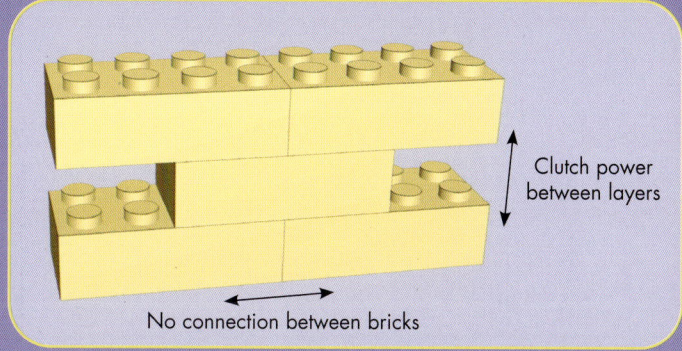

Clutch power between layers

No connection between bricks

With nothing but clutch power to cement the vertical joints between adjacent bricks, overlapping patterns like the staggered bond are much more important to LEGO builds than they are to real-world masonry, where adjacent, nonoverlapping columns of bricks can at least be bonded with mortar. Of course, it's theoretically possible to replicate mortar by gluing the pieces in a LEGO model together, but that would defeat one of the main advantages of LEGO as a creative medium—the ability to take your creations apart and rebuild them any number of times. In fact, the use of glue is generally frowned upon in the AFOL community. The only exceptions are for large LEGO models installed in public places, which typically have their pieces glued together for logistical and safety reasons.

ALTERNATING BRICK ORIENTATIONS

Sometimes you may need to build a LEGO wall that's more than 2 studs thick. In that case, the sturdiest way to build it is to alternate the orientation of bricks between odd and even layers of the wall. Say, for instance, the wall needs to be 4 studs thick. In one layer, you can place 2×4 bricks lengthwise in two rows (Figure 3-7, left). In the next layer, place 2×4 bricks crosswise (perpendicular to the previous layer), as shown on the right of Figure 3-7. This technique is based on the *English bond* pattern from masonry.

Figure 3-7: For thicker walls, one layer of bricks is placed lengthwise (left), and the next is placed crosswise (right).

Notice how the two rows of bricks in the first layer are offset from each other by 2 studs, much like a staggered bond pattern, so the joints aren't aligned between the two rows. Meanwhile, the crosswise layer of bricks straps the two rows in the first layer together, creating a fully interlocking structure. (Imagine if this layer were instead placed lengthwise, like the layer below—there'd be nothing holding the two rows of bricks together.) Offsetting the crosswise layer by 1 stud ensures that no vertical joints are aligned between the layers.

Alternating the orientation of bricks between layers can also be a useful technique for building walls that are 2 studs thick. Here, instead of using 2×4 bricks, you would have layers with two staggered rows of lengthwise 1×4 bricks alternating with layers of crosswise 1×2 bricks. It's no longer possible to offset the crosswise bricks in this variation of the English bond (each one is just 1 stud thick), so there will be some vertically aligned joints between the layers (see Figure 3-8). However, the aligned joints on the face of the wall are offset horizontally from the aligned joints on the back of the wall, thanks to the way the two rows of lengthwise 1×4 bricks are staggered. That way, the structure still interlocks.

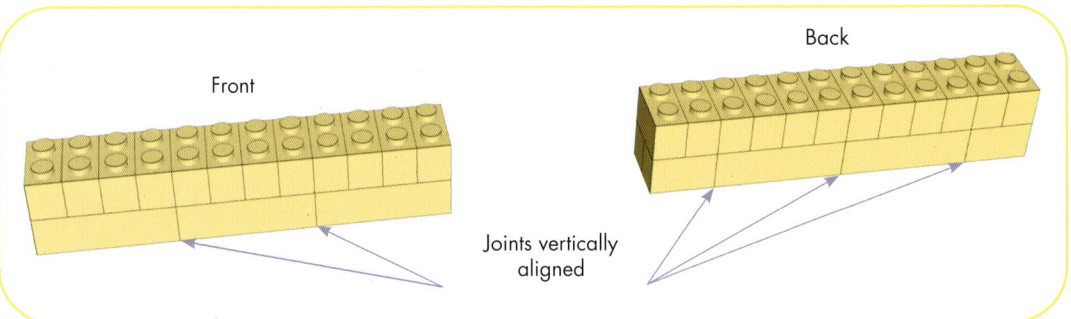

Figure 3-8: The front and back sides of a wall built using the English bond technique

REALIZING THE EMPIRE STATE BUILDING

We can use this modified English bond technique to begin realizing the 1:230 scale model of the Empire State Building we discussed in Chapter 2. In order to accommodate the pattern of window and wall segments on the building's facade, we'll need to work primarily with bricks that are 1 stud thick, for a total wall thickness of 2 studs. Alternating the orientation of bricks between the layers will be necessary to lock in the structure.

To recap, we established in Chapter 2 that each individual window or wall segment should be 1 stud wide, and that each floor of the model should be 2 bricks high. A close look at the actual Empire State Building indicates that the windows occupy about half the height of each floor, so we'll say that just the first layer of bricks in a floor (the odd layer) will feature windows; the second (even) layer will be a continuous horizontal wall.

LAYING OUT EACH FLOOR

Given that each window needs to be 1 stud wide and 1 brick high, we can't use dedicated LEGO window pieces like we did for the basic house from Chapter 2. We instead need to use regular bricks in a transparent color, preferably without bottom tubes for maximum transparency. Specifically, we'll use 1×2 trans-brown bricks placed crosswise so the windows will be transparent all the way through the wall. (This gives us the option to add internal lighting to the model if desired.) The wall segments between the windows in this layer can be made with 1×2 tan bricks, also placed crosswise, along with the occasional 1×1 or 2×2 brick at the corners. The full plan for the odd layer of each floor is shown on the left of Figure 3-9.

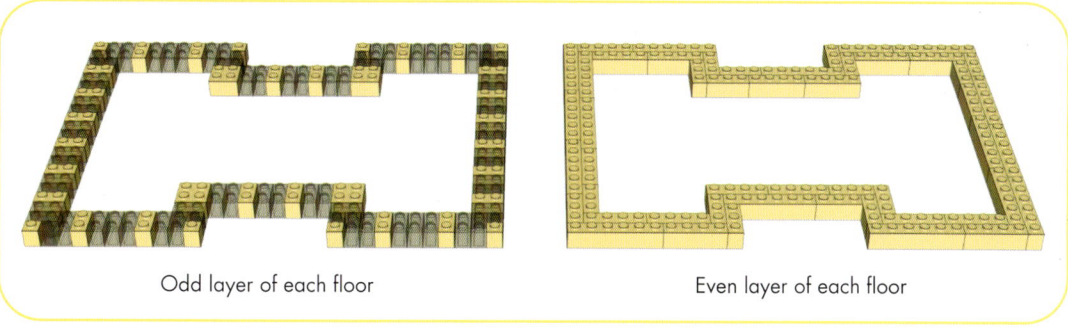

Odd layer of each floor   Even layer of each floor

Figure 3-9: The odd and even layers of each Empire State Building floor

In keeping with the alternating orientations technique, we'll design the even layer of each floor to have longer (1×3, 1×4, 1×6) tan bricks placed lengthwise in two rows, as shown on the right of Figure 3-9. Staggering the joints between the two rows ensures there won't be any vertically aligned joints running through the entire thickness of the wall.

GETTING THE WINDOWS RIGHT

We've created a plan for the two layers of each floor of the Empire State Building, but if we start stacking a few floors, we'll see that something doesn't look right (see Figure 3-10).

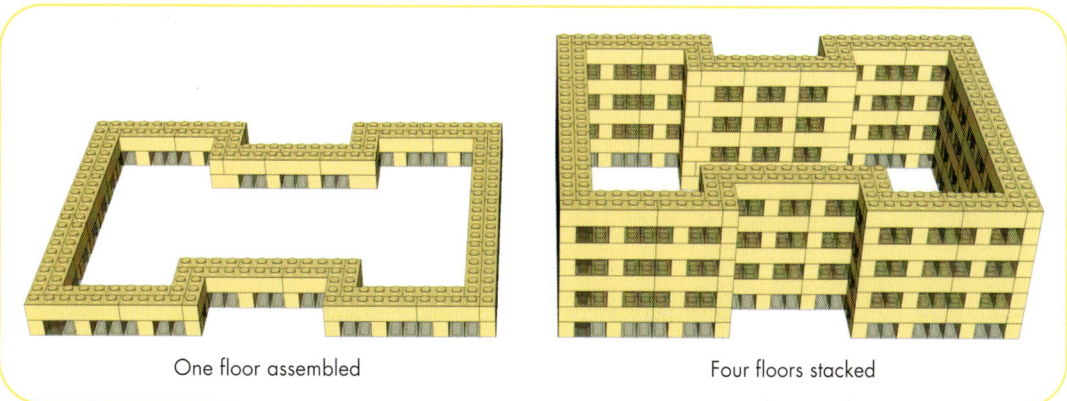

One floor assembled   Four floors stacked

Figure 3-10: Four floors stacked

43

In the actual Empire State Building, the windows appear to be in unbroken vertical lines. This effect is created by using gray accents above each window that blend in with the windows when viewed from a distance. An easy way to represent these accents would be to use 1×2 or 1×3 bricks of a different color—say, dark bluish gray—in the locations directly above each window, as in Figure 3-11. That way, when several floors are stacked, the darker window areas will form continuous vertical lines.

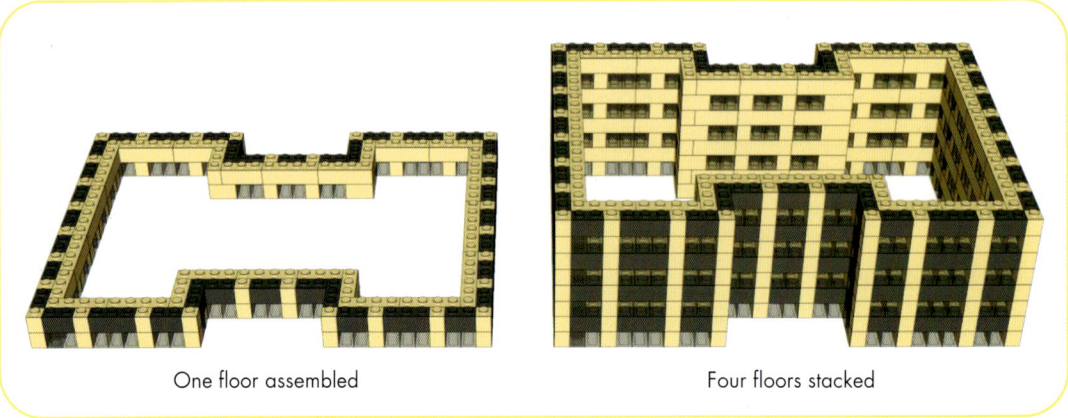

One floor assembled   Four floors stacked

Figure 3-11: Using dark bluish-gray accents above the windows on the even layers

OFFSETTING VERTICAL JOINTS

With the shorter dark bluish-gray pieces taking the place of longer tan pieces in the outer row of the non-window layers, there are now a lot of vertically aligned joints on the outer surface of the building. Thankfully, we still have the second (inner) row of lengthwise bricks to help strap everything together. The joints in this inner row should be arranged to align as little as possible with the joints in the outer row. But no matter what, we'll end up with a few places where the joints line up between the rows (see Figure 3-12).

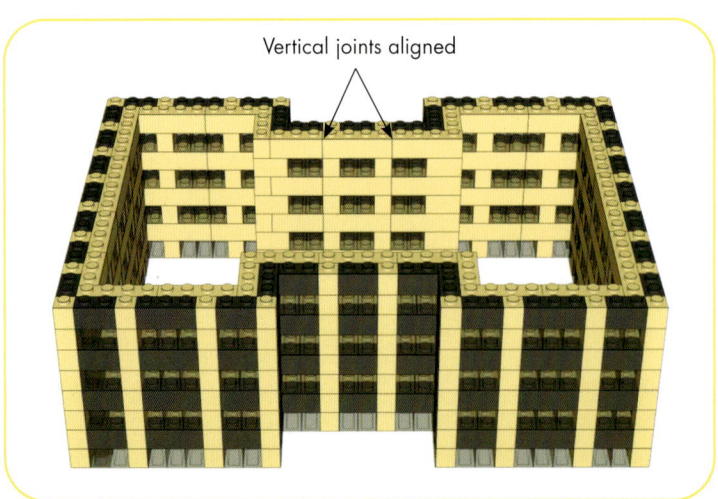

Vertical joints aligned

Figure 3-12: The vertical joints sometimes line up across multiple layers.

One way to work around this is to create two variants of the floor design. In one of the variants (say, the one used for the even-numbered floors), the inner row of bricks is offset by 1 stud horizontally compared to the other variant (used for the odd-numbered floors). The two variants are shown in Figure 3-13.

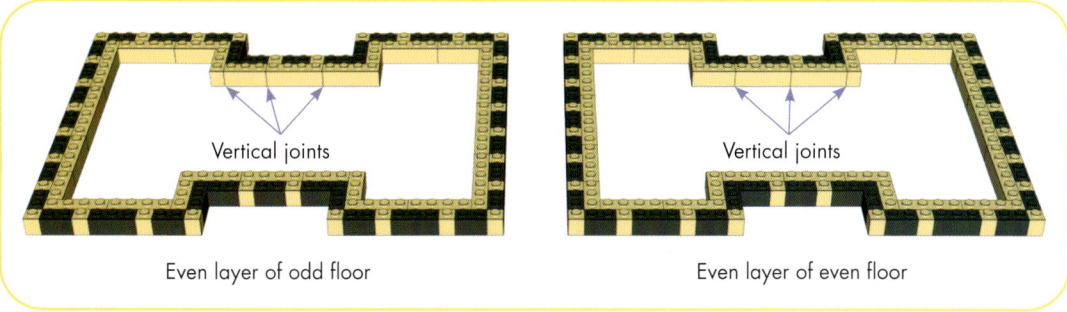

Figure 3-13: Two variants for odd and even floors

Once we start stacking up the floors, alternating the even and odd variants, we'll achieve a fully interlocked structure, where no vertical joint on the inner surface of the building aligns for more than three layers of bricks.

Alternating the orientation of bricks between odd and even layers of a model isn't just useful for creating the walls of a building. As we'll see in Chapter 9, it also helps when designing hollow LEGO sculptures.

INTERNAL SUPPORT STRUCTURES

Even with sturdy outer walls, some models also need an internal support structure for extra stability. Consider a building made by stacking multiple sections, each narrower than the one below it. An example is my model of the Hearst Tower in New York, which consists of a broad stone base topped by a narrower glass tower (see Figure 3-14). The footprint of the tower was too small for it to simply rest on the walls of the base. Instead, I had to build internal walls (shown in yellow) to support the weight of the tower.

There's a lot of leeway in building internal support structures. They're hidden from view, so you can use whatever types and colors of pieces you have available. You can even leave gaps in the support walls to save on bricks, as long as you use longer pieces that bridge these gaps and create an interlocked structure.

Internal supports can also help reinforce longer walls. Even when using the interlocking bond patterns we've discussed, long stretches of wall with no perpendicular cross members have a tendency to warp. In this case, some kind of inner bracing can increase structural integrity and keep the walls straight. Internal bracing is also important if you'll be moving a model around. Without extra support, the forces exerted on the model during the move may be enough to crush it.

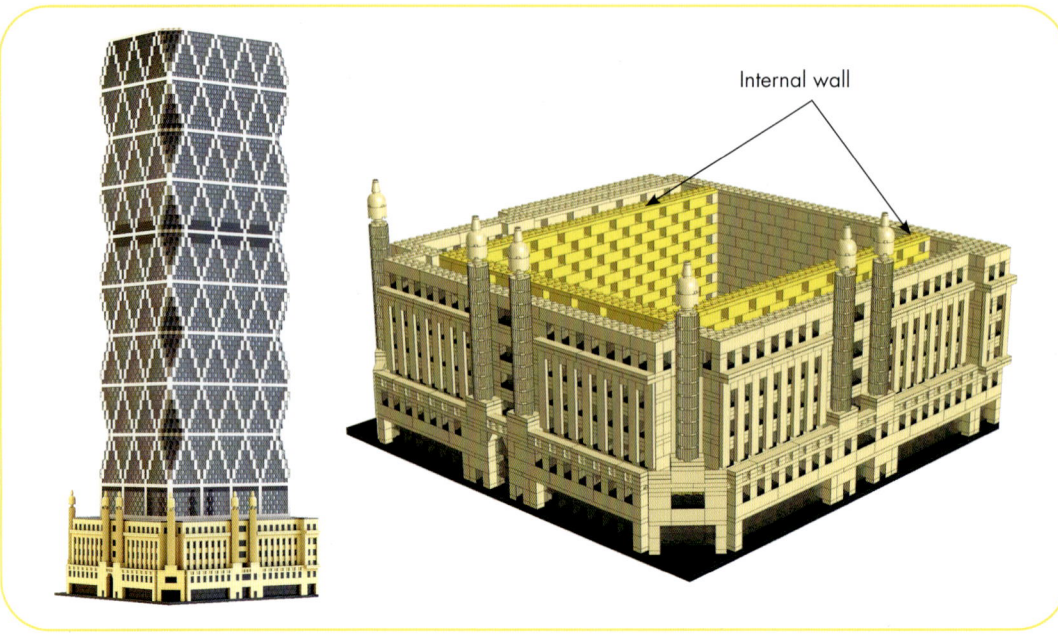

Figure 3-14: My model of the Hearst Tower with internal support walls

Internal bracing can take various forms. It can be made up of regular walls that act as cross members and bridge the gaps between parallel outer walls, or one or more columns of bricks that connect to the main outer walls via beams made from longer bricks or plates (see Figure 3-15, left). Many builders prefer to make the bracing from a network of Technic beams connected by Technic pins (Figure 3-15, right). Using Technic yields a structure that's just as strong as regular bricks and plates, but much lighter.

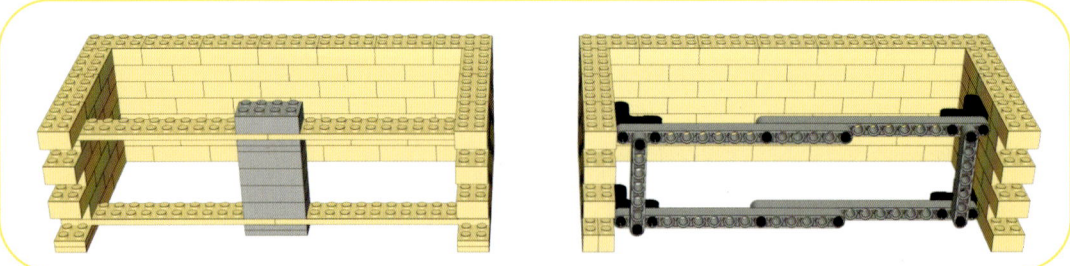

Figure 3-15: Internal bracing using a column of bricks (left) or a Technic frame (right)

Whatever approach you choose, this is one area where it helps to work with physical rather than digital bricks. Creating the right support structure usually requires some trial and error, and the feedback you get from experimenting with real LEGO pieces can't be matched in software.

CREATING SLOPED ROOFS

We've seen some techniques for building walls that rise vertically, but what about building surfaces that rise at an angle? For example, in Chapter 2, we left the model LEGO house without a roof. Let's revisit that model now and finish it. What's the best way to create its sloping roof using LEGO bricks?

WITH REGULAR BRICKS

Back in the earliest days of LEGO, when the system consisted of regular bricks and none of the specialized pieces that are available today, the only way to create a sloping roof was to stack bricks with a setback or offset of 1 stud on each successive layer. In fact, many of the LEGO sets released during this period featured stepped roofs built using this technique (see an example in Figure 3-16).

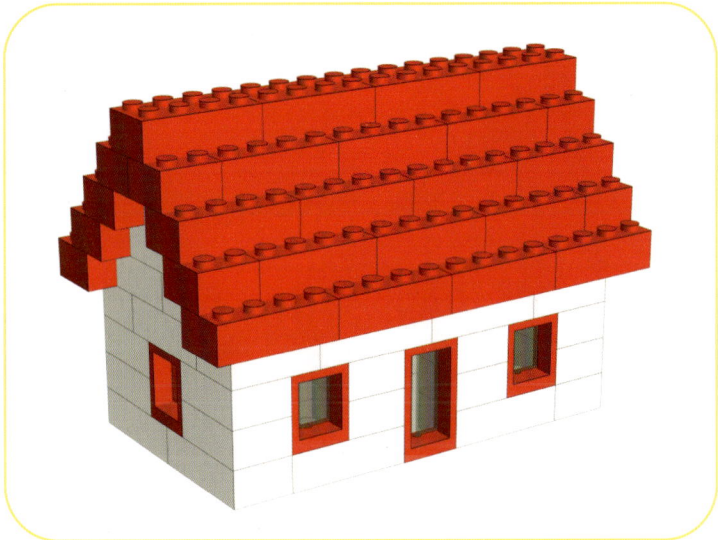

Figure 3-16: A stepped roof

Creating stepped roofs in this manner is only possible using bricks that are 2 studs deep. That way, the pieces that make up one layer of the roof can be attached to the inner row of studs on the layer below it. Figure 3-17 shows how the same method can be extended to all four sides to create a simple pyramid. The example in the figure tapers from a 16×16 base to a 2×2 apex.

Figure 3-17: A stepped pyramid

While stepped roofs work just fine, they aren't very realistic or elegant (especially with the exposed studs). If the shape of the pieces used for the stepped roof could be altered so that they create a continuous slope when stacked, we would be able to better represent the incline of a typical roof. This is exactly the idea behind LEGO's sloped roof bricks.

WITH SLOPE PIECES

The roof bricks that LEGO released in 1959 were among the earliest pieces designed for a specific purpose. The 2×4 roof brick (#3037) is like a regular 2×4 brick but with one row of studs removed and replaced with a slope (see Figure 3-18).

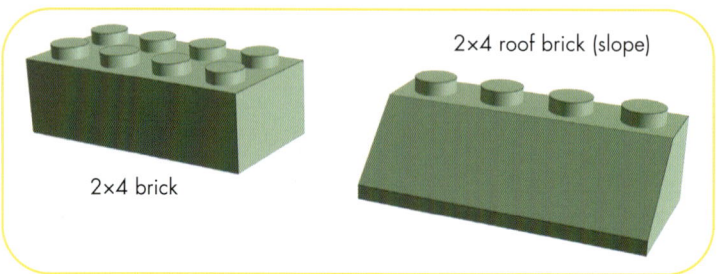

Figure 3-18: A 2×4 roof brick compared to a regular 2×4 brick

The slope traverses a vertical distance of 1 stud over a horizontal distance also equal to 1 stud, for an effective angle of 45 degrees. Since a LEGO brick is a little taller than 1 stud (3 plates versus 2.5 plates), this leaves a half-plate lip at the bottom of the slope (as shown in Figure 3-19). Because of this, the slope won't be perfectly smooth from one layer of bricks to the next, but in most cases the effect is close enough.

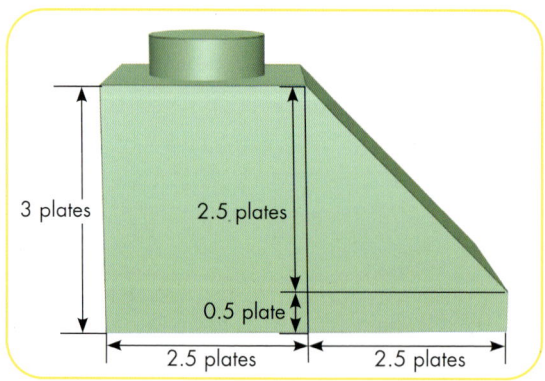

Figure 3-19: The dimensions of a roof brick

A VARIETY OF SLOPES

LEGO produces a variety of 45-degree roof pieces beyond the 2×4 slope. These include regular slopes in other lengths for the body of the roof, as well as pieces with slopes on multiple sides for the apex (top) of a roof or for inner and outer corners. Figure 3-20 shows a selection of 45-degree slope pieces.

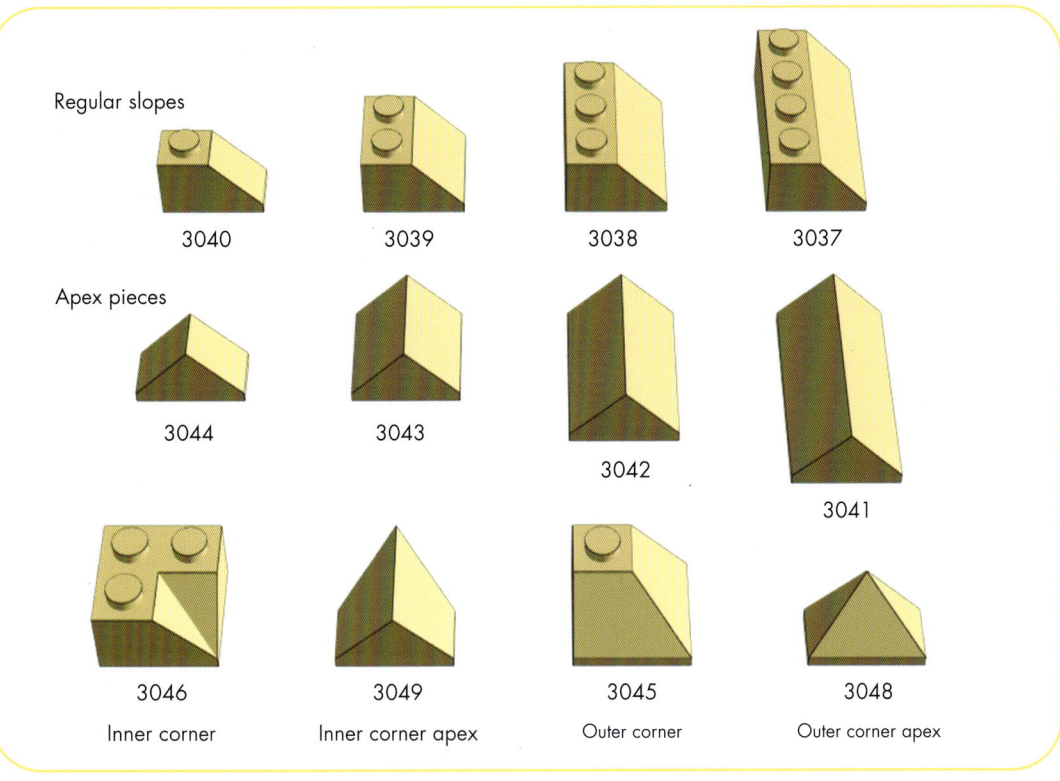

Regular slopes

3040   3039   3038   3037

Apex pieces

3044   3043

3042

3041

3046   3049   3045   3048

Inner corner   Inner corner apex   Outer corner   Outer corner apex

Figure 3-20: Different types of 45-degree roof pieces

Let's see how the pyramid from Figure 3-17 looks when we replace the regular bricks with roof bricks (Figure 3-21). The structure requires several different types of roof bricks, including bricks with slopes on two adjacent sides for the outer corners (#3045), as well as 1×2 apex bricks with slopes on three sides at the top (#3048).

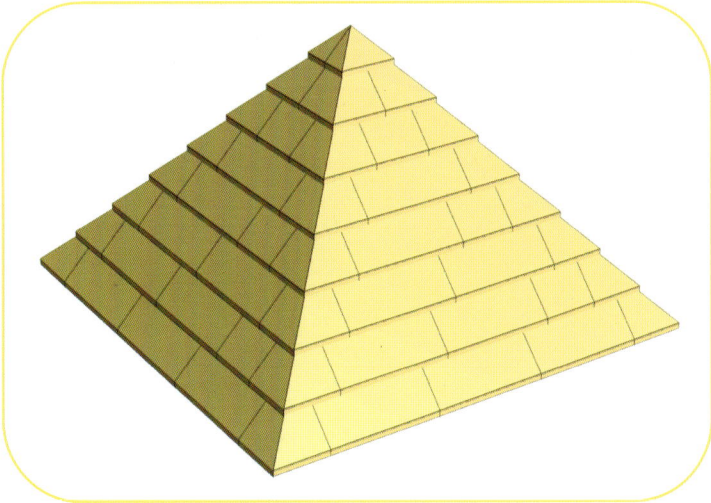

Figure 3-21: A pyramid built using roof bricks

Figure 3-22 shows a simple roof built with an even wider assortment of 45-degree roof pieces. Notice the use of inner corner (#3046) and inner corner apex (#3049) pieces, in addition to the other specialized roof pieces incorporated into the pyramid.

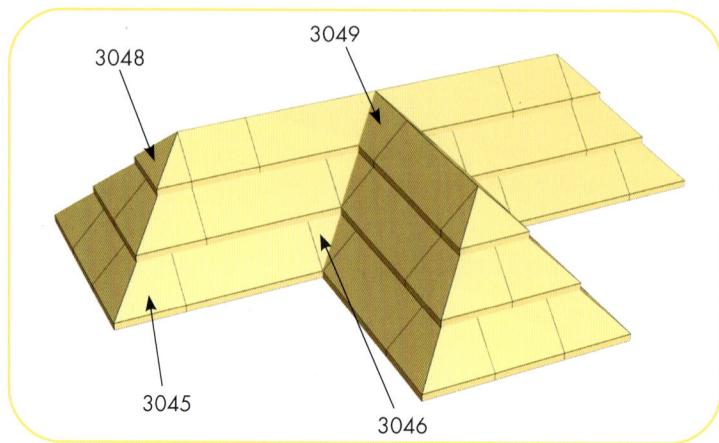

Figure 3-22: A simple roof built using various 45-degree roof pieces

A ROOF FOR THE MODEL HOUSE

Armed with these sloped pieces, we can build a roof for the model house. Let's start by building just a plain roof using mostly 2×4 roof bricks. Just as when you're building a regular wall with a staggered bond pattern, every other layer of slopes should be offset by 2 studs to create an interlocked structure. We can fill the gaps at the ends with shorter 2×2 roof bricks. When we get to the top of the roof, where the two sloped sides meet, we need to use 2×4 and 2×2 apex pieces with slopes on two sides. This is why we needed the depth of the house to be an even number of studs. The house is shown with the plain roof in Figure 3-23.

Figure 3-23: A plain roof without dormers

DORMERS

The original house we're trying to model featured three evenly spaced dormers. To add those, we need to create holes in the main roof where the dormers can be inserted. The dormers themselves can be built using regular bricks and window pieces (the same as what was used for the windows on the lower floors) and topped off with roof bricks (see Figure 3-24).

Dormer

Figure 3-24: Adding dormers to the roof

For the places where the dormer roof meets the main roof, we need to use inner corner roof pieces. Figure 3-25 shows the roof with all the dormers added.

Figure 3-25: The finished roof

If you take a close look at LEGO roof pieces, you'll see that the sloped portion isn't entirely smooth; it has a bit of a texture, originally intended to mimic the appearance of roof shingles. However, LEGO roof pieces have since found applications in more than just roofs and are now classified more generally as *slopes*. They can be used to create the outer shapes of everything from cars and boats to spaceships.

OTHER SLOPED ELEMENTS

We've been focusing on LEGO's original 45-degree roof slopes, but LEGO now produces bricks that slope at other angles as well. These slope pieces are broadly classified into families with names indicating the rough angle of the slopes. If you browse the online LEGO part catalog on BrickLink, you'll see slopes at 18, 30, 33, 45, 65, and 75 degrees. Here's a selection of them:

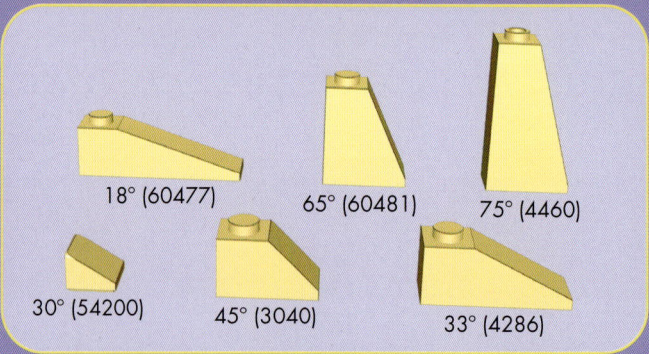

The 45-, 33-, and 18-degree slopes are each 1 brick high, with the sloped portion occupying 1, 2, and 3 stud widths, respectively. The sloped portion is 1 stud wide in the 65- and 75-degree slopes, but these pieces are 2 and 3 bricks high, respectively. The 30-degree slope, also called a *cheese slope*, is only 2 plates tall and has no stud on top. Regardless of the angle, all LEGO slopes have a half-plate lip at the bottom. The reason is attributed to limitations in the injection molding process used to manufacture LEGO bricks.

The LEGO catalog also features *inverted slopes*, which have studs on top and a slope on the bottom, alongside the anti-studs. Inverted slopes can be used to create arches in buildings, or they can be combined with regular slopes and other elements to create the outer shapes of cars, ships, aircraft, and other vehicles. Here are some examples:

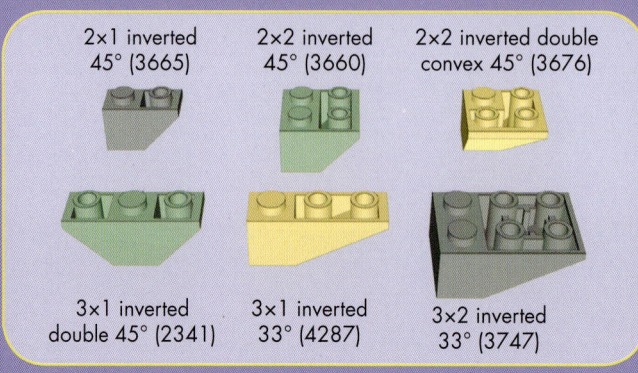

Finally, LEGO also produces elements that have curved rather than straight slopes, as shown in the following figure.

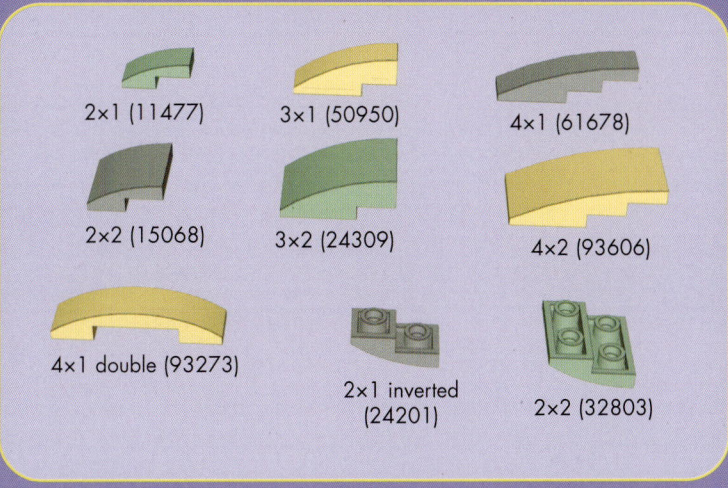

2×1 (11477) 3×1 (50950) 4×1 (61678)

2×2 (15068) 3×2 (24309) 4×2 (93606)

4×1 double (93273) 2×1 inverted (24201) 2×2 (32803)

Curved slopes come in both regular and inverted variants. They give you even more options for sculpting the surfaces of your models.

ILLEGAL TECHNIQUES

The term *illegal techniques* may suggest that some ways of building LEGO models run afoul of an official law governing the use of LEGO pieces. In fact, no such law exists; *illegal* instead indicates that a technique uses LEGO bricks in a way that wasn't intended by the manufacturer. While some of these techniques may be quite harmless, others can put undue strain on the LEGO pieces involved and permanently damage them.

PONY EARS

An example of a harmless illegal technique is the *pony-ear technique*, where a plate or tile is wedged vertically in the gap between studs on a brick or a plate (see Figure 3-26).

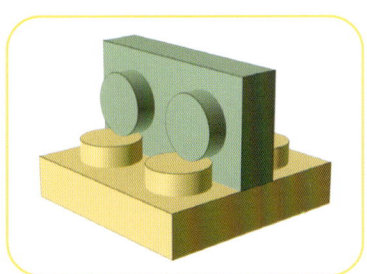

Figure 3-26: The pony-ear technique

This technique was commonly used in official LEGO sets before LEGO started making pieces that enabled sideways building, or SNOT (which we'll cover in Chapter 5). It takes advantage of the fact that the gap between adjacent studs on a brick or a plate is 0.32 cm, which is exactly equal to the thickness of a plate or a tile. Also, the studs are inset 0.16 cm from the edge of a plate (which is equal to the height of the stud itself), so theoretically there's no chance of a collision between the studs when you place a plate or tile vertically in the gap between two rows of studs. In practice, however, the raised "LEGO" text on the top of a stud can cause a mild collision when a plate is wedged in the gap between studs. The pony-ear technique is therefore considered illegal except when a tile is used (and the tile version of this technique continues to be used sparingly in official sets).

DIGITAL BUILDING TIP

Joining pieces together using illegal techniques can be easy with physical bricks, but it can be a frustrating exercise when you're building digitally in Studio. With the pony-ear technique, for example, the side of a tile doesn't have a legal connection to the gap between the studs on a plate, so the pieces won't snap together automatically. Instead, you'll have to position the tile manually. First, select the tile and rotate it by 90 degrees (using the arrow keys on your keyboard) so that it's perpendicular to the plate you want to "attach" it to. Then use the W, A, S, and D keys to move the tile up, down, and sideways into position. Before you do that, be sure to click the **Grid** button in the toolbar and select the **Fine Grid** option. This will allow you to move pieces in the smallest increments possible (1 LDU). It is also possible to use the Move tool to move pieces. You can select the **Move** tool in the toolbar or right-click the piece and select **Move** from the drop-down menu. This will bring up three arrows, each aligned along a different axis. You can click one of these arrows and drag your mouse to move the piece along its axis. The Move tool also brings up a text field showing the x, y, and z coordinates of the piece (expressed in LDUs). The coordinates in this text field can be edited to position the piece very precisely.

54

STUD REVERSAL

Other common illegal techniques include ones that can be used to reverse the direction of studs in a LEGO build (we'll look at legal ways of doing this in Chapter 5). These techniques involve wedging pieces like cheese slopes into the undersides of bricks or plates and attaching them to other bricks or plates that have their studs facing the other way. This creates a combination "brick" that has studs on both the top and bottom (which can be useful in certain builds), as shown in Figure 3-27.

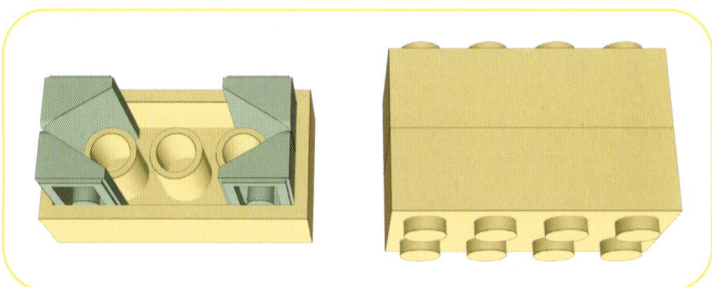

Figure 3-27: Stud reversal

TECHNIC BRICKS WITH HOLES

Some illegal techniques use Technic bricks with holes. These holes are intended to hold Technic pins, but they can also accommodate LEGO studs, albeit with a very tight fit. This makes it possible to mix regular LEGO bricks with Technic bricks in ways that may not always be legal. For example, Figure 3-28 shows a 1×1 round plate attached to a 1×1 Technic brick with a hole.

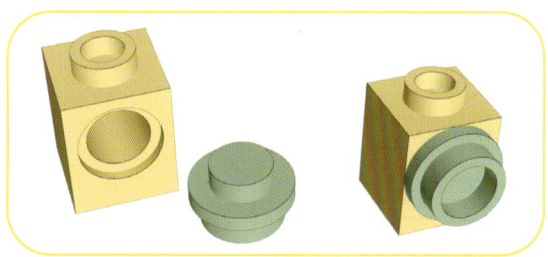

Figure 3-28: A plate attached to a Technic brick with a hole

We'll revisit this technique in Chapter 5, where it will prove to be very handy on our 1:230 model of the Empire State Building.

SUMMARY

In this chapter, we discussed a few basic techniques for creating sturdy LEGO builds, such as using a staggered bond pattern and alternating the orientation of bricks between successive layers of a model. We also reviewed the different types of LEGO slope pieces and how they can be used. We'll continue our exploration of LEGO building techniques in Chapter 4, where we'll look at ways to create subtle detail using half-stud offsets.

PART II

BREAKING FREE FROM THE GRID

We normally build models by stacking bricks one on top of another, while staying within the confines of the LEGO system's regular square grid of stud locations. Finding creative ways to break free from this grid is the key to designing interesting LEGO models. Keep reading to learn about techniques for offsetting bricks by fractions of the grid unit, building sideways, creating angled walls, and even approximating round shapes.

HALF-STUD OFFSETS

When you're building with LEGO, you can only place each piece such that its studs (bumps) line up with the studs on the layer immediately below it. If you start with a 32×32 baseplate, for example, that means you're restricted to a 32×32 grid of possible locations (separated by increments of 1 stud) where you can place the pieces in your next layer. This can be quite limiting. Say, for example, you're building a wall and you want a section to be recessed. The smallest amount you can normally set back the recessed section is 1 stud, since the pieces have to align with the grid determined by the layer below.

But what if you want to create a more subtle effect? Would it be possible to set the recessed section of the wall back by half a stud instead of a full stud? Yes! Using pieces called *jumper plates*, you can create a *half-stud offset* between one layer and the next. In this chapter, we'll look at jumper plates and half-stud offsets, and explore the different ways you can use them in LEGO models.

TYPES OF JUMPER PLATES

There are four main varieties of jumper plates, shown in Figure 4-1. They all feature open-style studs on top due to the nature of LEGO's injection molding process.

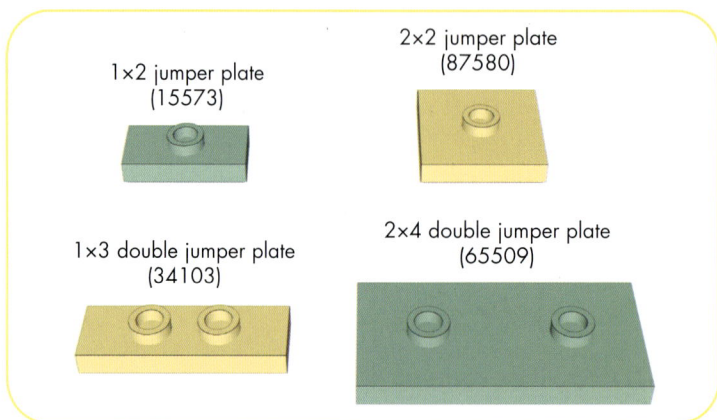

Figure 4-1: Four types of jumper plates

THE 1×2 JUMPER

The standard type is the 1×2 jumper plate (#15573), pictured in the top left of Figure 4-1. It's the same size as a regular 1×2 plate, but it has just one open stud located exactly halfway between where the two studs on a regular 1×2 plate would sit.

When you attach a brick or plate on top of a 1×2 jumper plate, it gets shifted, or offset, by half a stud relative to the LEGO grid below. In this way, jumper plates let you set a wall section back by half a stud instead of a full stud, as seen in Figure 4-2. On the left, I've recessed the 2×2 brick by a full stud so it still aligns with the main LEGO grid. On the right, I've used jumper plates to recess the 2×2 brick by just half a stud. It no longer aligns with the main grid, and as a result, the back of the 2×2 brick has to rest on a 1×2 tile; there aren't any studs for the back half of the brick to attach to.

A half-stud offset creates a more subtle effect than a full-stud offset. Using 1×2 jumper plates, you can add features that either are inset or protrude slightly relative to the main wall of a building, as we'll explore in this chapter. This can give a LEGO facade more dimension and texture.

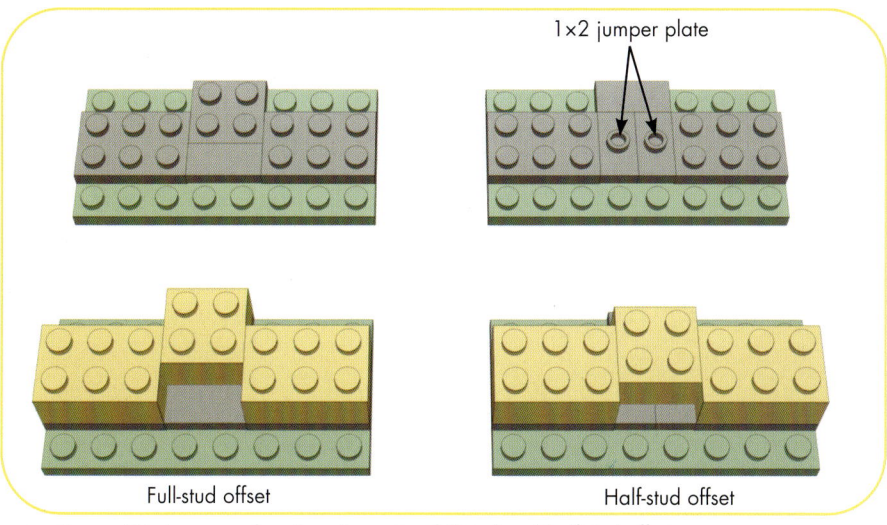

1×2 jumper plate

Full-stud offset | Half-stud offset

Figure 4-2: A recessed wall section using full-stud and half-stud offsets

THE 2×2 JUMPER

While a 1×2 jumper plate allows you to create a half-stud offset in one dimension (either front to back or sideways), you can do the same in two dimensions using a 2×2 jumper plate (#87580). This jumper plate (see Figure 4-1, top right) has a single stud exactly in the center relative to the four studs on a regular 2×2 plate. Figure 4-3 compares full-stud and half-stud offsets in one and two dimensions, using the 1×2 and 2×2 jumper plates, respectively.

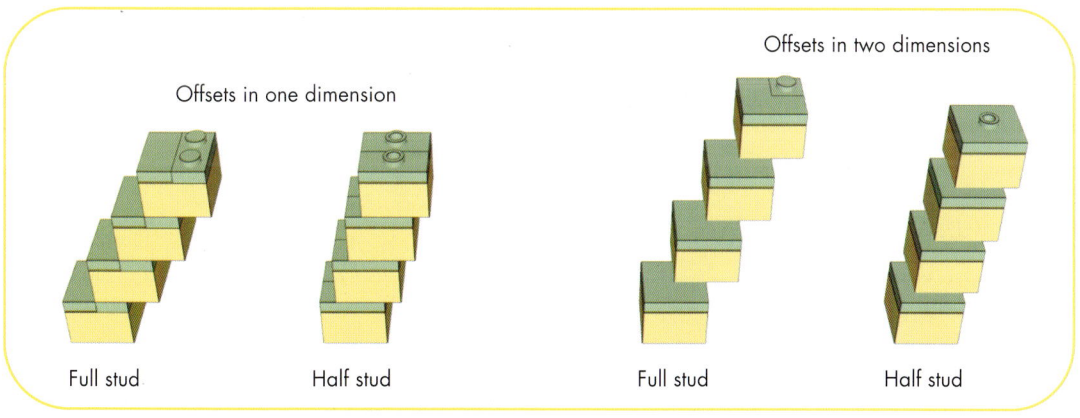

Offsets in two dimensions

Offsets in one dimension

Full stud | Half stud | Full stud | Half stud

Figure 4-3: Offsets in one and two dimensions

Notice on the left that the 1×2 jumper plates create a sideways offset, but each layer is still aligned front to back. On the right, the 2×2 jumper plates create a half-stud offset in both dimensions. In both cases, the successive layers of half-stud offsets create a smoother taper than is possible with full-stud offsets, allowing you to create buildings that gracefully get narrower as they rise. This is another application of jumper plates that we'll examine in this chapter.

DOUBLE JUMPERS

LEGO has recently expanded its catalog to include "double jumper" counterparts for its 1×2 and 2×2 jumper plates. These are the 1×3 jumper plate with two studs (#34103) and the 2×4 jumper plate with two studs (#65509), shown on the bottom row of Figure 4-1. The 2×4 jumper plate is nothing more than two 2×2 jumper plates joined together, but the 1×3 jumper plate can be useful in its own right. Consider a variation on the example from Figure 4-2, where the recessed portion of the wall is only 1 stud wide (Figure 4-4). Using a 1×3 jumper plate instead of a 1×2 provides two offset studs available as connection points instead of one, allowing the recessed 1×2 brick to be attached more securely.

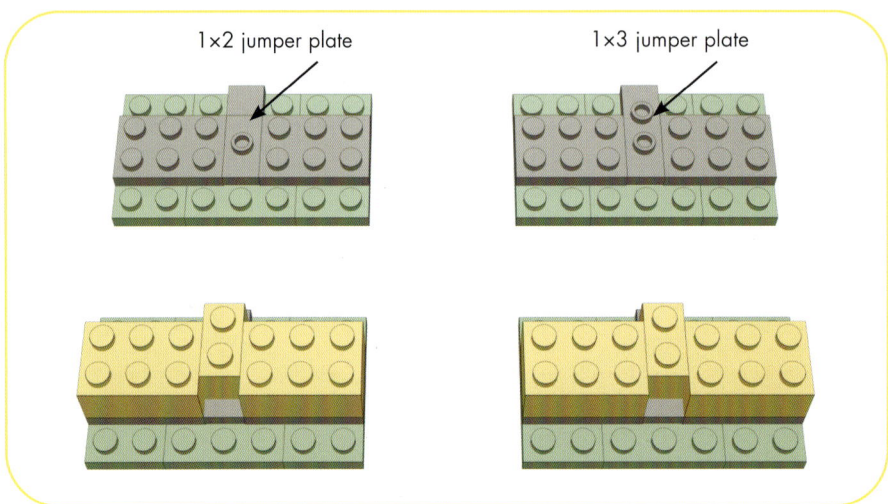

1×2 jumper plate   1×3 jumper plate

Figure 4-4: Using a double jumper plate for a more secure connection

Because jumper plates have fewer studs on top than they do anti-studs underneath, they have more clutch power on the bottom than on top. This makes it harder to remove them using the regular end of a brick separator. But much like tiles, jumper plates have grooves on their bottom edge to provide some leverage for prying them off.

COMMON JUMPER PLATE APPLICATIONS

Jumper plates and the half-stud offsets they produce have many applications in LEGO builds. You can use them to add surface texture to a structure, or to recess sections of a wall, windows, or other architectural features. Or you can use them to center elements that are an odd number of studs wide relative to elements that are an even number of studs wide. You can also use them to create structures that taper smoothly. In this section, we'll discuss these common applications for jumper plates and look at examples of each.

ADDING SURFACE TEXTURE

When we last left our simple model house in Chapter 3, one of the details we had omitted was the shutters on either side of each window. These shutters need to stick out slightly from the walls of the house, adding visual interest and surface texture to the structure. There was no easy way to create this effect using regular bricks. But with jumper plates and half-stud offsets, creating shutters that stick out will be easy.

To imitate the texture of real shutters, we'll use special 1×2 bricks with a fluted profile (#2877), stacking four of these bricks to match the height of the windows. Then we can attach a shutter on either side of each window using 1×2 jumper plates. First, a pair of jumper plates at the bottom offsets the shutter outward by half a stud. Then, a second set of jumper plates at the top reverses the half-stud offset to get us back to the original alignment of the wall (see Figure 4-5).

Figure 4-5: A shutter assembly created using jumper plates

We'll need to remove some of the existing wall bricks on either side of the windows or replace them with shorter ones to be able to fit the new shutter assemblies. Once the assemblies are in place, they'll leave gaps on the top and bottom that are too thin for bricks and will need to be filled in using plates. Figure 4-6 shows the model with the shutters added.

Figure 4-6: The model of the house with shutters added

I've also included a front door and a covered front stoop topped with sloped roof pieces in the final version of the model.

RECESSING WALLS, WINDOWS, AND OTHER FEATURES

You've already seen how 1×2 jumper plates can add subtle detail to a building by setting back sections of a wall by half a stud. You can use the same technique to recess windows or other architectural features of a building, setting them back slightly from the main surface of a wall. Figure 4-7 shows one way to recess a window.

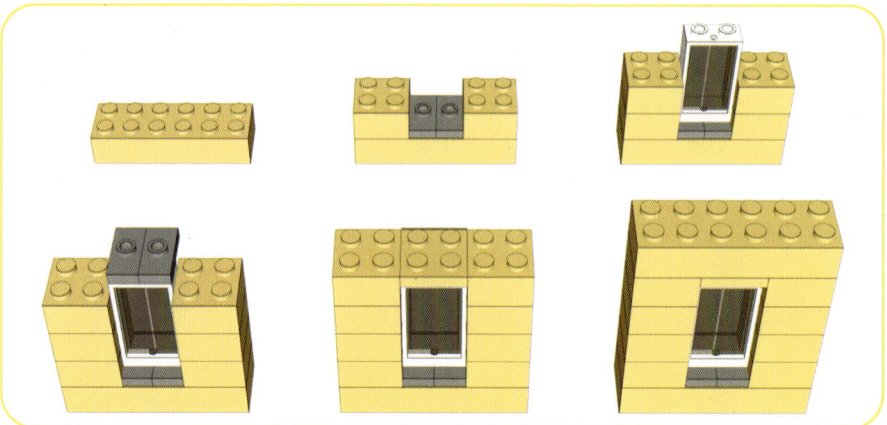

Figure 4-7: A recessed window created using half-stud offsets

The window sits on top of two 1×2 jumper plates, recessing it by half a stud. Similarly to the shutters example, I've used a second set of jumper plates on top of the window to get back to the normal alignment of the studs for the upper portion of the wall.

In my own models of skyscrapers, I use half-stud offsets extensively to create recessed effects. In the case of my model of the Empire State Building, for example (discussed in Chapters 2 and 3), I used jumper plates to create recessed wall sections at the top of the building, as shown in Figure 4-8. I placed jumper plates above the five central columns of windows, allowing the wall sections above them to be slightly recessed. The recessing also makes room for the decorative elements above the middle three windows.

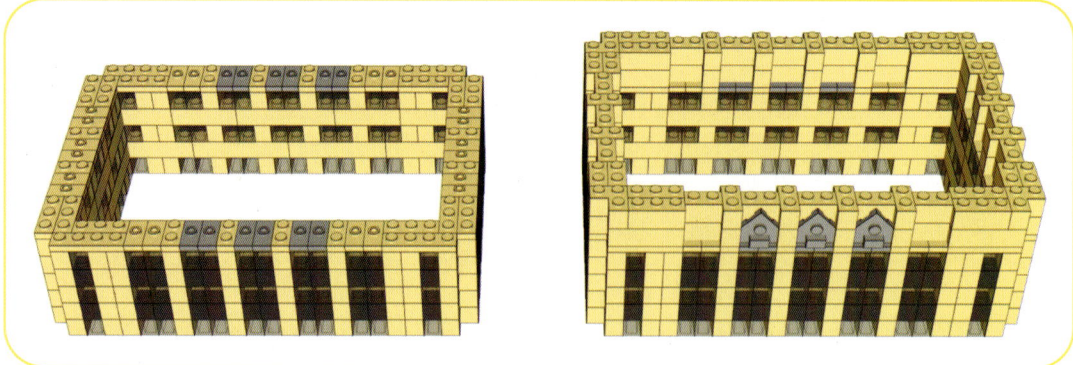

Figure 4-8: The recessed wall section at the top of the Empire State Building

Another iconic skyscraper built around the same time as the Empire State Building is 70 Pine Street in New York. Like most skyscrapers from this period, 70 Pine Street has a distinctive tapered shape; it rises through a complex series of setbacks to a white-accented crown.

To accurately represent this crown, my model of 70 Pine Street recesses the windows themselves using jumper plates (see Figure 4-9). A lower layer of jumper plates (white) offsets the windows. Then a second set of jumper plates (tan) further sets back the section above the windows as the top of the building tapers up to the spire.

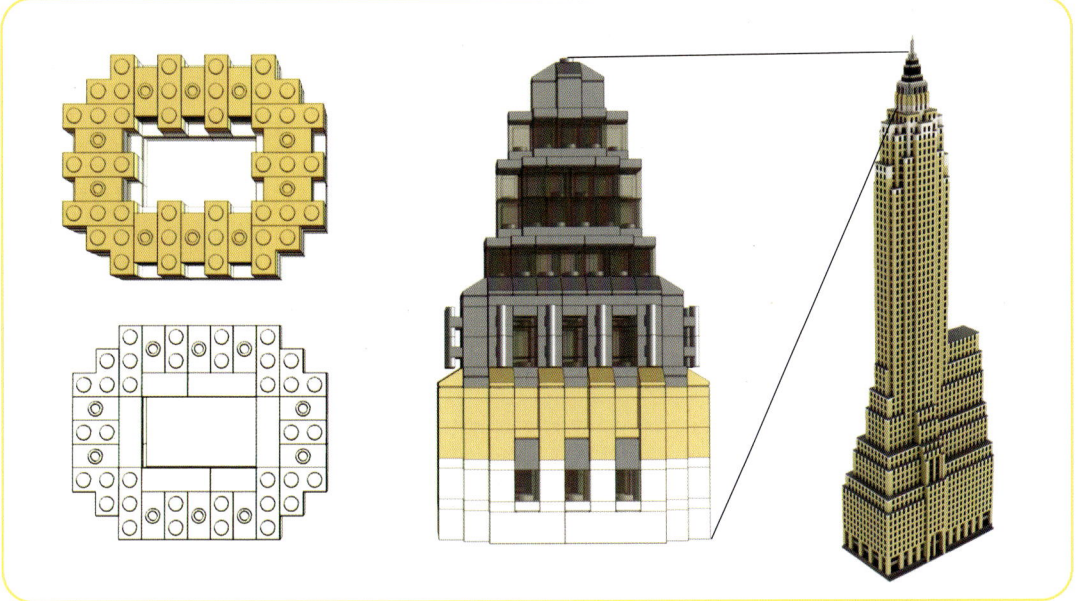

Figure 4-9: The recessed windows at the top of 70 Pine

CENTERING ELEMENTS

Another great application for jumper plates is centering an element with an odd number of studs relative to something with an even number of studs, or vice versa. For example, the top portion of my Empire State Building model has windows that are 1 stud wide, but they need to be centered relative to the windows below them, which are 2 studs wide. Figure 4-10 shows how I placed the 1×1 trans-brown bricks representing the narrower windows on top of 1×2 jumper plates in order to center them relative to the wider windows below.

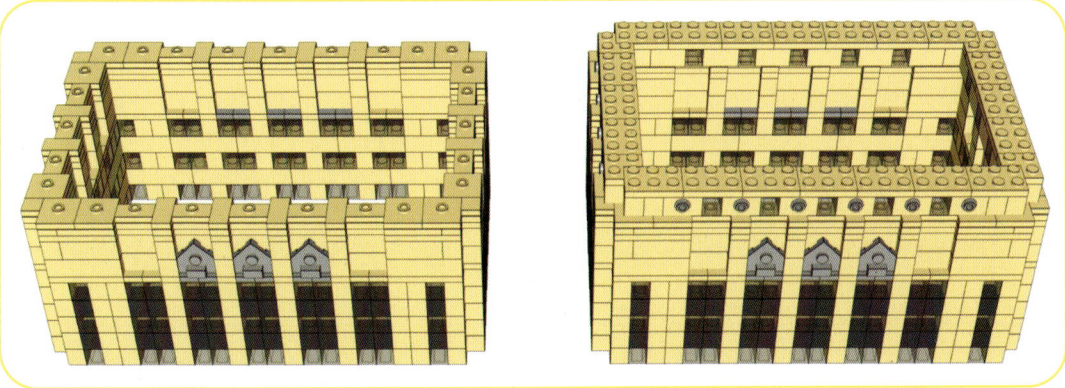

Figure 4-10: Centering windows at the top of the Empire State Building

The Blue Mosque in Istanbul is considered one of the crown jewels of Ottoman architecture. With its elegant composition of ascending domes and six slender, soaring minarets, it dominates the city's skyline. To correctly represent the windows on the sides of the building in my LEGO model, I needed to fit three smaller arched window openings inside a bigger arch.

I used 1×3 arch pieces (#4490) for the smaller arches, but unfortunately, once you go above 3 studs wide, the only arch pieces that are available span an even number of studs (4, 6, 8, and so on). Figure 4-11 shows my solution.

Figure 4-11: Centering small arches inside a bigger arch using jumper plates

I used a 1×8×2 arch (#3308) for the larger opening. Behind it, I used jumper plates to center the three smaller arches, which together have a total span of 7 studs, relative to the larger arch's 8-stud span. Additional jumper plates on top of the smaller arches, including a 1×3 jumper plate, bring the grid back into alignment, allowing me to cap the whole assembly with a 2×8 plate.

SMOOTHING OUT TAPERS

As buildings rise, jumper plates help achieve smoother tapers than you can achieve with regular bricks and plates. As mentioned, these tapers can unfold evenly in both the left-to-right and front-to-back dimensions, thanks to the 2×2 jumper plate. Sometimes, you only need to use this taper effect for certain sections of a building. For example, like many of the skyscrapers built during the early 1930s, the Empire State Building and 70 Pine Street have top sections that taper as they lead up to their spires. I used 2×2 jumper plates for the tapers in my models of these buildings. Figure 4-12 shows an example from my Empire State Building model.

The base of the Empire State Building's spire was built in multiple layers, each topped with 2×2 jumper plates. This creates the subtle stepped effect at the top of the model, as the building narrows to its spire.

Whereas the Empire State Building required tapering using half-stud offsets only at the top, near its spire, other buildings taper evenly over the whole course of their height, which requires

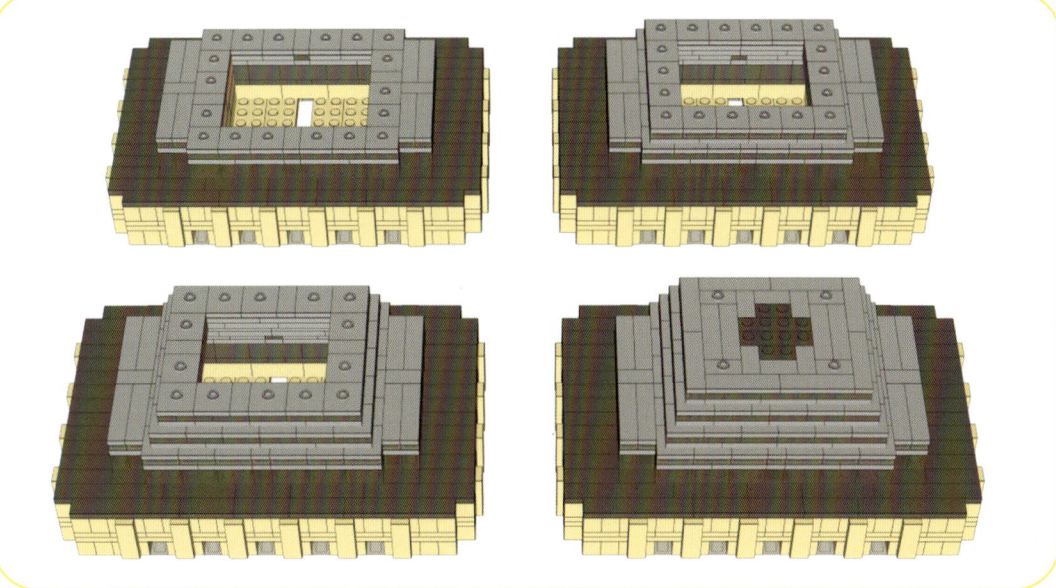

Figure 4-12: Tapering at the top of the Empire State Building

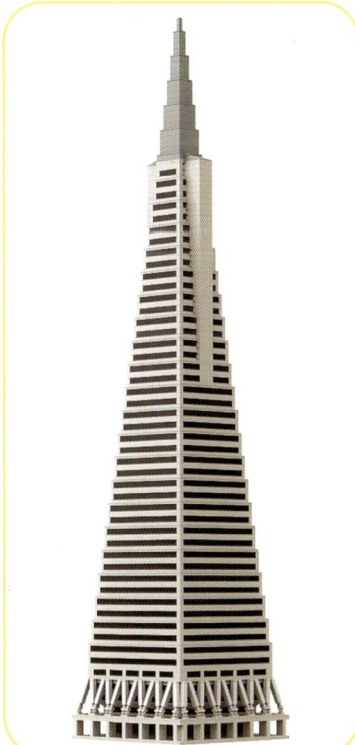

Figure 4-13: A model of the Transamerica Pyramid in San Francisco

much more extensive use of half-stud offsets. For example, the Transamerica Pyramid in San Francisco is a skyscraper with a square base that gradually tapers to a point at the top. Based on the scale I was using for my model (shown in Figure 4-13), I needed the building to taper from 28 studs per side to 7 studs per side over the course of 42 floors. With regular bricks and plates, the smallest amount I would have been able to taper the model is 2 studs (1 stud on each side) approximately every four floors. Using 2×2 jumper plates, however, I was able to taper the model by just 1 stud (half stud on each side) every two floors, minimizing the jaggedness of the model and producing a significantly smoother taper.

My Transamerica Pyramid model reveals a downside to half-stud offsets: they can leave small gaps in a building. While the main part of the Transamerica Pyramid tapers from bottom to top, the building is flanked by "wings" on two sides, structures that hold an elevator shaft and stairwell and that rise vertically, without any tapering. To create these wings, I needed to have normal wall sections intersecting with the tapered walls. However, because of the half-stud offsets, the tapered walls don't form a straight vertical line at the point where they meet the wings. Instead, they zigzag back and forth, as shown on the left side of Figure 4-14, leaving a half-stud gap between the pyramid walls and the wings every other set of two floors.

Unfortunately, these gaps are unavoidable. As you can see on the right side of Figure 4-14, I did my best to plug the gaps by attaching tiles to the wings.

Figure 4-14: A close-up of half-stud gaps between tapered walls and the wings of the Transamerica Pyramid

TAPERING BY UNEQUAL AMOUNTS

Not all tapered buildings taper at the same rate on all sides. The John Hancock Center in Chicago pioneered the use of a tubular system in skyscraper construction. The building's rectangular tube consists of a grid of beams and columns on its perimeter with diagonal braces on the exterior walls, giving the tube extra strength. The tapered shape was intended to provide additional structural stability against wind forces.

To accurately represent the proportions of this building, I needed to taper the long sides of my model at a faster rate than the short sides. Specifically, the long sides taper by 1 stud (half a stud on each side) every six floors, while the short sides taper by 1 stud every eight floors. Figure 4-15 shows the complete model.

To achieve this effect, I couldn't simply use 2×2 jumper plates, as I did on the Transamerica Pyramid, since they would produce an even taper on all sides. Instead, I mostly used 1×2 jumper plates, oriented two different ways—one way for long-side tapers and the other way for short-side tapers. I could use 2×2 jumpers only on floors where both sides tapered at the same time.

Figure 4-16 shows a simplified example of a 12-story building that tapers from 8×6 studs at the base to 3×3 studs at the top. The longer sides go from 8 studs to 3 studs in six stages, so they need to be tapered by a stud every two floors. Meanwhile, the shorter sides go from 6 studs to 3 studs in four stages, so they need to be tapered by 1 stud every three floors.

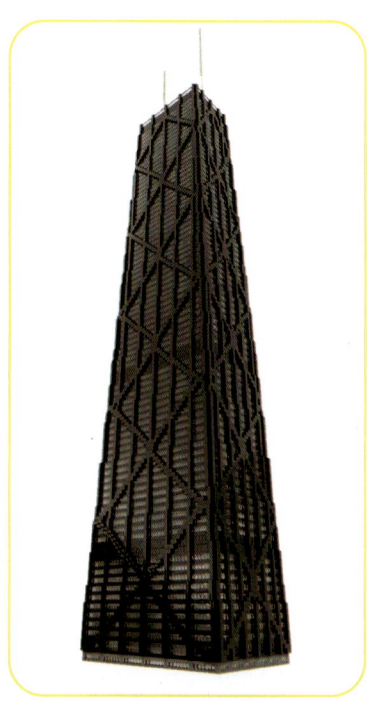

Figure 4-15: A model of the John Hancock Center in Chicago

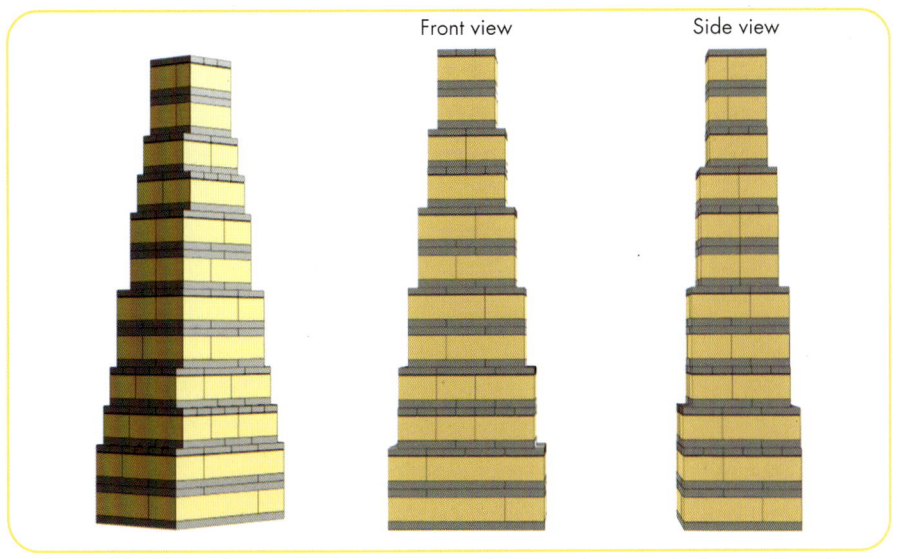

Figure 4-16: A simple example showing tapering by unequal amounts in the two dimensions

The tapers may seem a little haphazard at first glance (especially at this small scale), but they start to make more sense when you look at the front and side views of the building. Count out the two stories of tan bricks between tapers when viewed from the front, and the three stories of tan bricks between tapers when viewed from the side, and you'll see how each axis of the building (x and y) tapers evenly, at its own rate. Figure 4-17 shows a breakdown of the offsets needed at each floor to achieve this tapering.

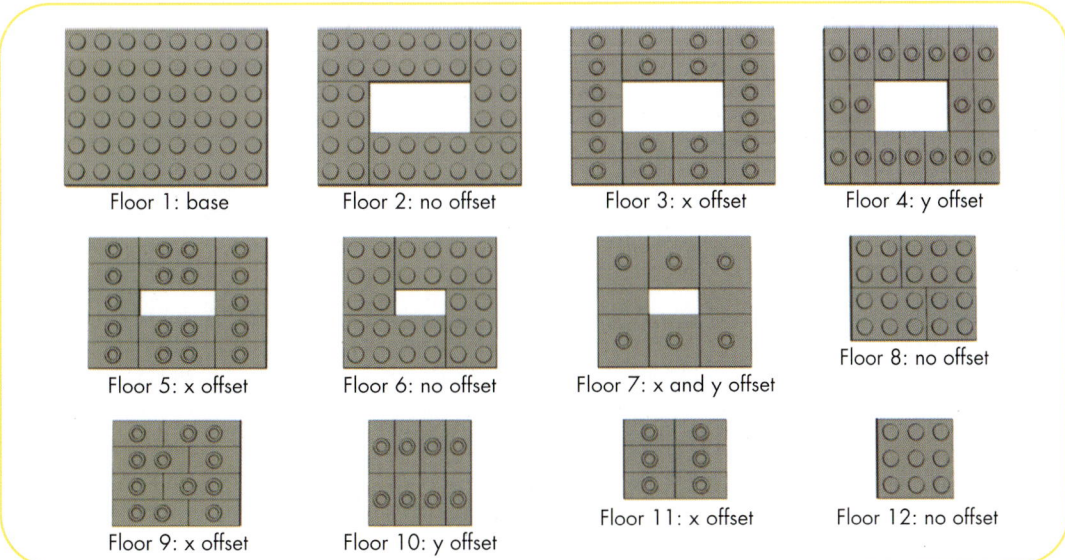

Figure 4-17: A breakdown of offsets needed at each floor in our simple example

On floors where the building tapers in the x direction, 1×2 jumper plates oriented horizontally create the offset. (Some 1×3 jumper plates are mixed in as well.) On floors where the building tapers in the y direction, the 1×2 jumper plates are oriented the other way. At the sixth floor, the building tapers in both directions, so 2×2 jumper plates can be used.

USING JUMPER PLATES WITHOUT AN OFFSET

While jumper plates are mostly used for creating half-stud offsets, you can also attach plates and bricks on top of jumper plates without any offset at all, as shown in Figure 4-18. This connection takes advantage of the fact that the open stud on top of a 1×2 jumper plate can fit perfectly with the tube on the underside of a 1×2 plate.

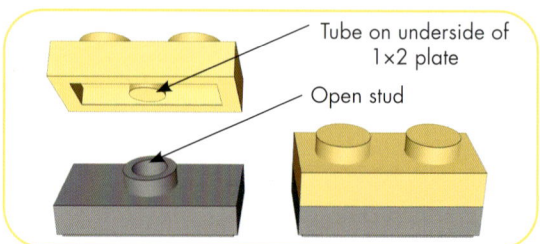

Figure 4-18: The tube on the underside of a regular plate can fit into the open stud on the top of a jumper plate.

Because jumper plates have higher clutch power on the bottom than they do on the top, they tend to stay firmly attached to the layer below while allowing the layer above to be separated without much effort. This type of connection is therefore very useful in big models that have multiple sections that need to be taken apart and put back together easily. I tend to mix in jumper plates with tiles at the seams between the different sections that make up my skyscraper models, for instance, and I've never had to worry about loose pieces coming off while taking the sections apart. Figure 4-19 shows an example from my Empire State Building model.

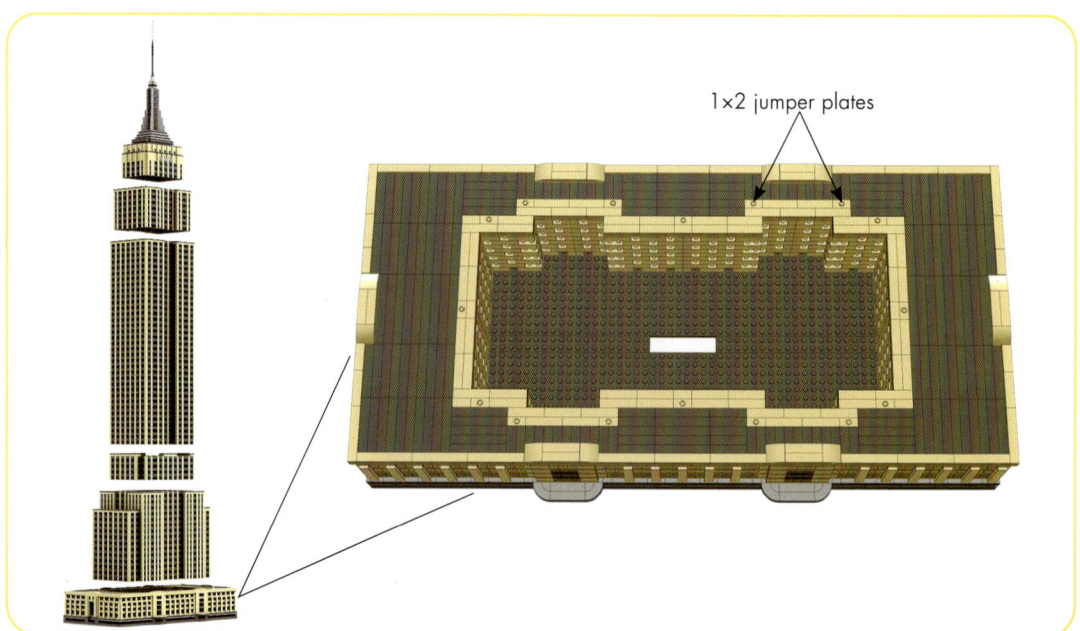

Figure 4-19: Jumper plates used at the seams between different sections of the Empire State Building

On the right side of the figure, you can see the seam between the base of the model and the next section above it. Notice the sprinkling of jumper plates among the tiles. The jumpers hold the section above securely in place while still allowing it to be separated from the base easily.

HALF-STUD OFFSETS WITHOUT JUMPER PLATES

You don't always need jumper plates to create half-stud offsets. Sometimes you can create them simply by using the right combination of elements with open studs and 1×*n* bricks or plates that have bottom tubes that fit in these open studs. Figure 4-20 shows an example.

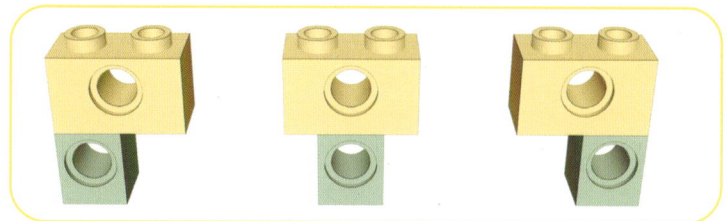

Figure 4-20: Creating a half-stud offset using the open stud on the top of a 1×1 Technic brick

Technic bricks have open studs on top and narrow tubes underneath. Therefore, they can be stacked to align either with the normal LEGO grid (as on the left and right of the figure) or with a half-stud offset (as in the middle of the figure).

SUMMARY

The next time you build something with LEGO, don't underestimate those lowly jumper plates. They may be small and easy to overlook, but as you've seen in this chapter, they're a powerful design tool thanks to their half-stud offset. Using offset techniques, you can add subtle details like recessing and tapering to your models, making them more realistic. We'll next look at some techniques to attach LEGO pieces sideways when we explore SNOT in Chapter 5.

SIDEWAYS BUILDING (SNOT)

All the LEGO techniques we've seen so far have involved stacking bricks one on top of the other, with their studs pointing up. But if you look through the bricks that make up any recent LEGO set, there's a good chance you'll see some that have studs not just on top but on their sides as well, as shown in Figure 5-1. These bricks let you get away from the typical studs-on-top orientation and attach LEGO pieces sideways.

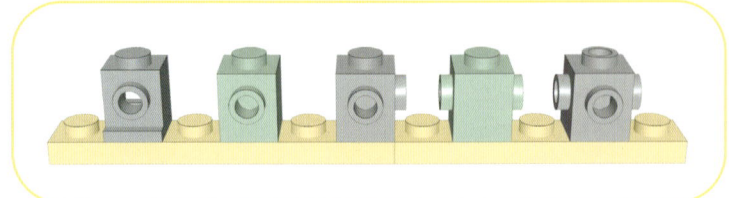

Figure 5-1: Bricks with studs on their sides

Sideways building techniques are collectively known as *SNOT*, a not-so-elegant acronym for "Studs Not On Top." SNOT is a relatively recent development in the overall history of LEGO, with most SNOT elements being introduced in the 2010s and later, but SNOT is a powerful addition to your LEGO building technique arsenal. By building sideways, you can take full advantage of the geometry and proportions of LEGO elements and create shapes and details that wouldn't be possible to achieve through simple stacking.

In this chapter, we'll revisit the geometry of LEGO bricks and see how it comes into play when you start building sideways. We'll also cover some of the common types of LEGO elements designed for sideways building and look at some more specialized SNOT techniques, including how to use half-plate offsets to seamlessly integrate the horizontal and vertical portions of a model.

SNOT GEOMETRY

Recall from Chapter 1 that a basic 1×1 brick is a little taller than it is wide. Its height (0.96 cm) is equivalent to 3 plates, while its width (0.8 cm), which we usually refer to as 1 stud, is equivalent to 2.5 plates. The resulting height-to-width ratio of 6:5 is important to remember when you're building sideways. To understand why, consider a 1×6 plate. If you stand it up vertically, as shown on the left of Figure 5-2, it comes to 6 studs or 6 × 2.5 plates = 15 plates tall. That's the same as the height of a wall built by stacking five 1×1 bricks on top of each other, since 3 plates high per brick × 5 bricks = 15 plates. So 6 studs is equivalent to 5 bricks tall.

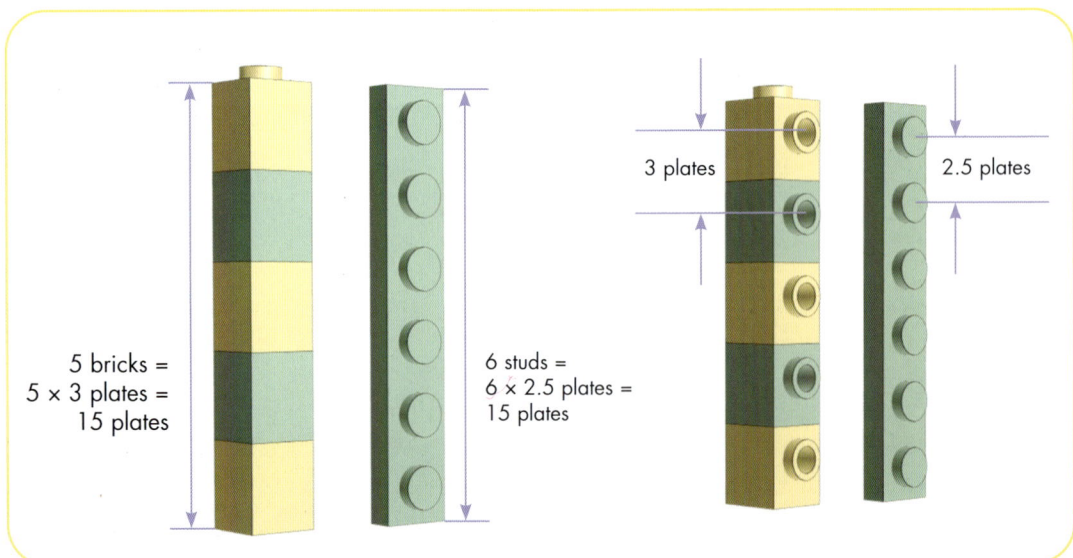

3 plates

2.5 plates

5 bricks =
5 × 3 plates =
15 plates

6 studs =
6 × 2.5 plates =
15 plates

Figure 5-2: Five brick heights are equivalent to 6 studs, but the stud locations don't line up.

If the wall used 1×1 bricks with studs on one side (#87087), could we just attach the 1×6 plate to the face of the wall? Not really. The spacing between the studs on the wall face would be equal to the height of each brick, which is 3 plates. This wouldn't match the spacing between the 6 studs on the 1×6 plate, which are 1 stud = 2.5 plates apart (see the right half of Figure 5-2). The studs just don't line up.

To solve this problem, first try attaching a 1×1 plate to the face of a 1×1 SNOT brick (see Figure 5-3, left). Its upper edge will be flush with the top of the brick (not including the stud), but a small sliver at the base of the brick will be left exposed. The brick is 3 plates high, and the plate attached to it is 2.5 plates tall, so that small sliver is 3 − 2.5 = 0.5 plates high.

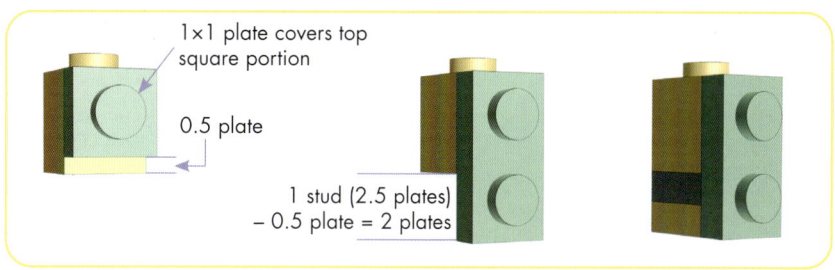

Figure 5-3: A stack with a brick and two plates is 2 studs tall.

Now replace the 1×1 plate with a 1×2 plate (Figure 5-3, center). This adds 1 stud (2.5 plates) of length to the sideways plate, enough to cover the 0.5-plate sliver plus an extra 2.5 − 0.5 = 2 plates of overhang. If you now add two 1×1 plates below the SNOT brick, the height of the stack (1 brick + 2 plates) will match the width of the 1×2 plate (Figure 5-3, right). This illustrates a way to get the studs on the face of a wall to line up with the studs (every other stud, in this case) on the brick or plate we're attaching sideways; simply sandwich two layers of plates between each layer of SNOT bricks (see Figure 5-4, left).

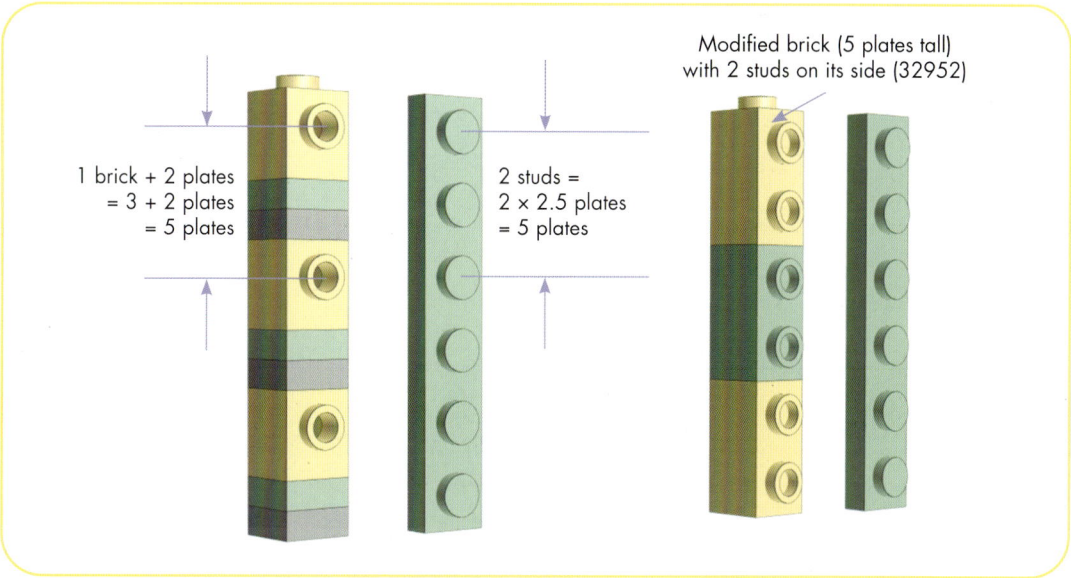

Figure 5-4: Sandwiching two plates between the layers of bricks gets the studs to line up correctly. Another option is to use SNOT bricks that are each 5 plates tall.

A relatively recent addition to the LEGO catalog even allows us to have a full complement of studs on the face of the wall. There's a special SNOT brick (#32952) that's 5 plates tall and has two rows of studs on its side. The right half of Figure 5-4 shows what the wall would look like using this newer SNOT brick. There's an attachment point for each of the sideways plate's 6 studs.

TYPES OF SNOT ELEMENTS

Now that we've seen how the geometry of LEGO bricks affects sideways building, let's take a closer look at the most common types of SNOT elements. These include headlight bricks, bricks with studs on their sides, plates with studs on their sides, and brackets. We'll examine the features of these elements and consider some ways they can be used. But this isn't an exhaustive survey of SNOT elements. There are other sideways building techniques that use Technic or specialized elements like plates with clips, lamp holders, and so on. The techniques we cover here will be enough to get you started on your sideways journey.

HEADLIGHT BRICKS

LEGO sets going back to the early 1970s included sideways building in some rudimentary form—including some techniques, like the pony-ear technique we covered in Chapter 3, that would be considered illegal now (see Figure 5-5, left). This was done only in a very limited fashion, however, and purely for decorative purposes. Sideways building didn't come into its own until the first LEGO brick with a stud on its side was introduced in 1980: the headlight brick (#4070).

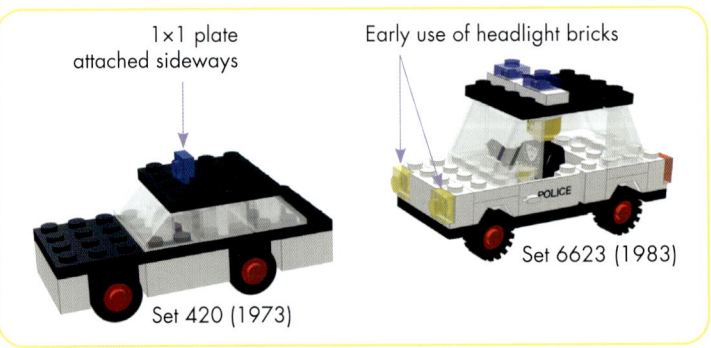

Figure 5-5: Set 420 with a 1×1 plate attached using the pony-ear technique, and set 6623 with an early use of headlight bricks

The arrival of the headlight brick seemed quite unremarkable at the time. LEGO designers had no good way of attaching headlights to the cars and other vehicles that were a part of the Classic Town sets being released around then. LEGO designer Erling Dideriksen invented the headlight brick (also known as the Erling brick) to solve this problem (see Figure 5-5, right). It's basically a 1×1 brick with a stud on one of its sides (in addition to the stud on the top).

Interestingly, the stud on the side of the brick is recessed by half a plate. In fact, the entire top square portion of the face of the brick is recessed, and so that top portion is only 2 plates deep. Meanwhile, the bottom sliver is intact, creating a notch that's half a plate high, as shown in Figure 5-6. Presumably this was done to ensure that the "headlight" plate wouldn't stick out too much.

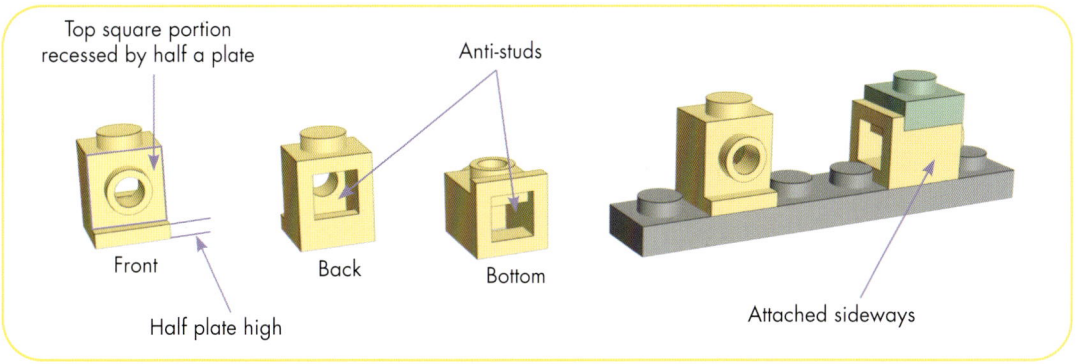

Figure 5-6: The geometry of a headlight brick and the two ways it can be attached

Even more interesting is the fact that the headlight brick has a square cutout on the back, essentially forming an anti-stud. This allows it to be attached like a regular brick, even when it's tipped over on its back (see Figure 5-6, right). This is probably an indication that Dideriksen had more uses in mind for this brick than its nickname would suggest.

More than 40 years later, headlight bricks remain one of the most versatile LEGO elements. Given their unique geometry, there are quite a few interesting ways to join headlight bricks together. Figure 5-7 shows some examples.

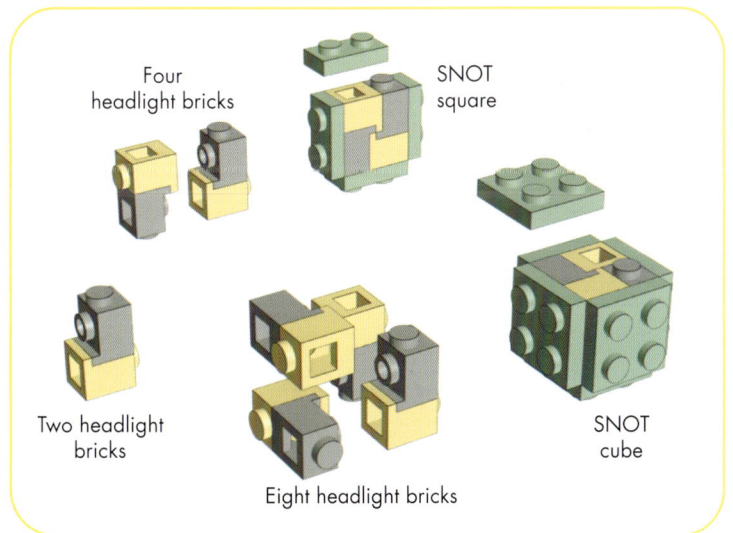

Figure 5-7: A SNOT square and cube created using headlight bricks

Using four interlocking headlight bricks and 1×2 plates, you can create a SNOT square that has studs on four sides. Similarly, with eight headlight bricks and some 2×2 plates, you can create a SNOT cube with studs on all six sides.

Figure 5-8 shows another interesting headlight brick configuration. Connect six of them together and you get the logo for the now retired LEGO Creator Expert theme.

We'll revisit headlight bricks later in the chapter when we explore the various offsets they can produce.

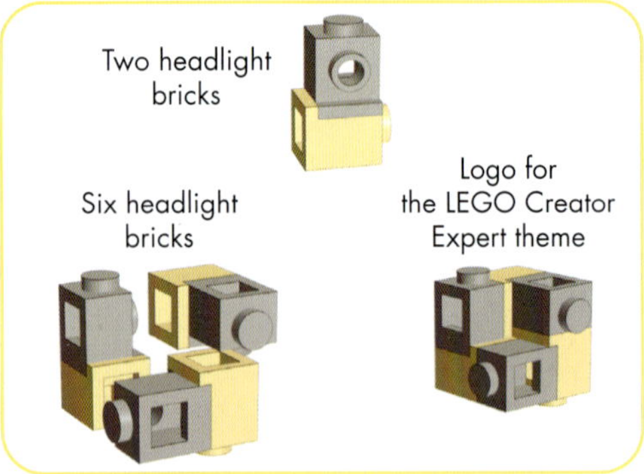

Two headlight bricks

Logo for the LEGO Creator Expert theme

Six headlight bricks

Figure 5-8: Using headlight bricks to create the logo for the LEGO Creator Expert theme

BRICKS WITH STUDS ON THEIR SIDES

Since the introduction of the headlight brick, LEGO has produced a wide range of SNOT bricks with studs on their sides (see Figure 5-9), though it took a surprisingly long time to get there. In fact, the regular 1×1 brick with one stud on its side, probably the most fundamental SNOT element, didn't make an appearance until 2009. However, the 1×1 brick with studs on all four sides (#4733) has been around a lot longer (since 1985).

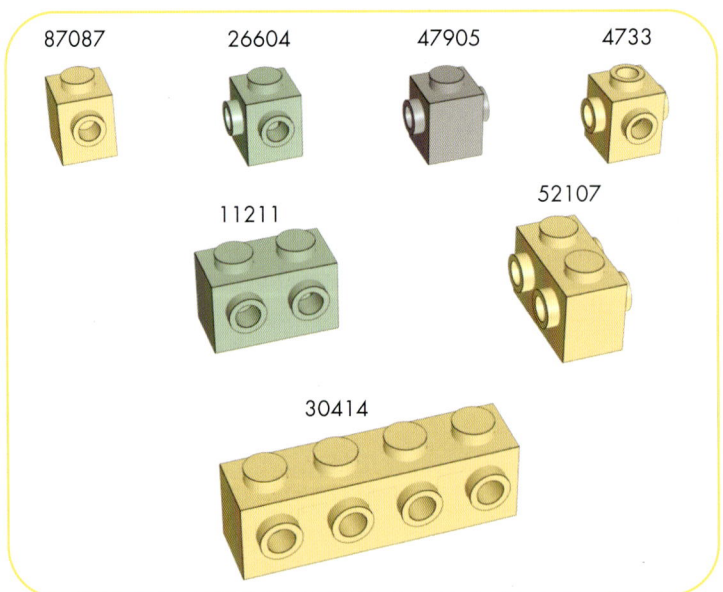

87087   26604   47905   4733

11211   52107

30414

Figure 5-9: Bricks with studs on their sides

Rounding out the family of 1×1 SNOT bricks are the bricks with studs on two opposite sides (#47905, introduced in 2004) and two adjacent sides (#26604, introduced in 2017). There are larger 1×2 and 1×4 counterparts of some of these SNOT bricks as well (#11211, #52107, and #30414), along with some relatively new bricks that were designed with SNOT geometry in mind. These new bricks (see Figure 5-10) are 5 plates tall with two rows of studs on their sides (#32952, #22885, #67329, and #80796).

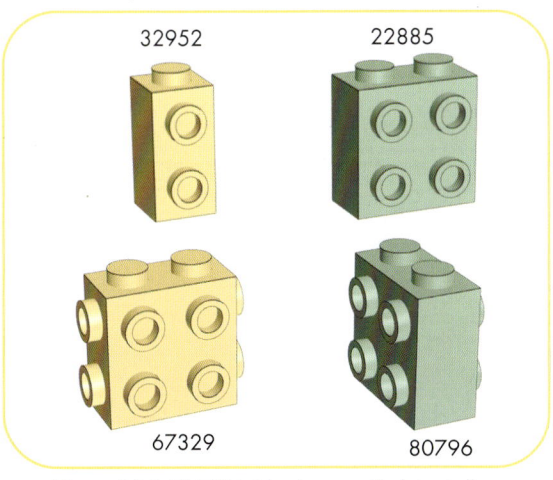

Figure 5-10: SNOT bricks that are 5 plates tall with studs on their sides

DETAILING WALLS

SNOT bricks are great for adding details to the external walls of buildings. I used them in the base section of my Empire State Building model, for example, to recess the windows slightly compared to the walls. Instead of pushing the windows in, I pushed the rest of the wall out by attaching sideways 1×6 tiles to the wall sections between the windows (see Figure 5-11).

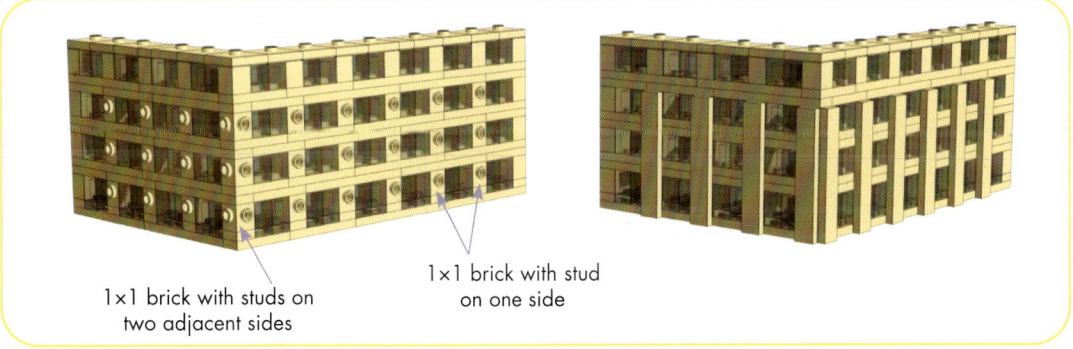

1×1 brick with studs on two adjacent sides

1×1 brick with stud on one side

Figure 5-11: SNOT used to create a recessed window effect on the base of the Empire State Building

Each floor in this section of the model was 5 plates high, perfect for SNOT. I sandwiched each layer of bricks between two layers of plates to correctly space the sideways studs. I needed 1×1 bricks with two studs on adjacent sides in the corners and 1×1 bricks with a single sideways stud everywhere else.

It was a little trickier to create the same effect on the base of my model of New York's Hearst Tower. Here the scale was bigger (1:156 rather than 1:230), calling for 7 plates per floor. To be able to attach 1×8 tiles to the faces of the wall, I needed to somehow get the studs on the face of the wall to be 5 plates apart, even if each floor was 7 plates high. As you can see in Figure 5-12, I was able to mix and match bricks and plates, sometimes within the same layer. This created a fully interlocked structure with the right vertical spacing between both the windows (7 plates) and the studs on the face of the wall (5 plates).

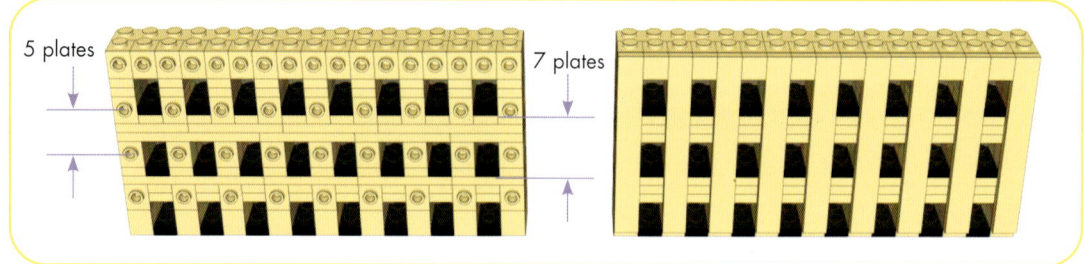

Figure 5-12: SNOT used to create a recessed window effect on the base of the Hearst Tower

BUILDING SPIRES AND COLUMNS

Bricks with studs on all four sides were used extensively in the past, when the other types of 1×1 bricks with studs on their sides didn't exist. In most of those cases, we can now get away with using the newer bricks with one or two studs on their sides. Applications that specifically require the studs on all four sides are limited, at least in architectural builds. The bricks can possibly be used in spires on the top of a building or to create octagonal columns, as shown in Figure 5-13.

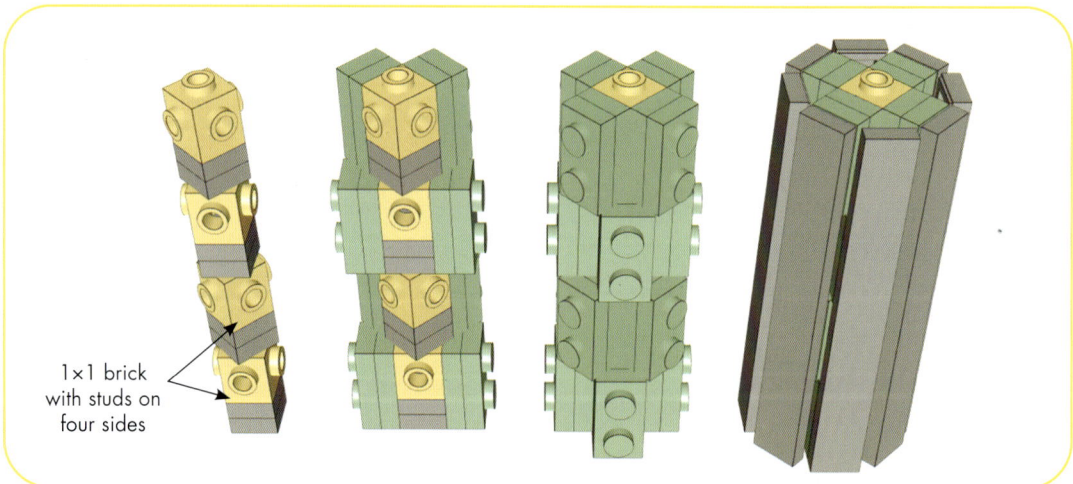

1×1 brick with studs on four sides

Figure 5-13: An octagonal column created using 1×1 bricks with studs on all four sides

To achieve the octagonal effect, rotate every other layer of SNOT bricks 45 degrees, producing studs in eight directions. Then build out horizontally from the SNOT bricks with two layers of 1×2 plates to even out the surface before covering up the interior with long tiles.

CREATING SNOT CORES

You can also combine bricks with studs on one side and two adjacent sides to create *SNOT cores*, hidden central structures with studs in many directions for attaching plates and curved slopes. These cores can form the basis for cylinders, spheres, and other shapes (see Figure 5-14). We'll look at this technique in more detail in Chapter 7.

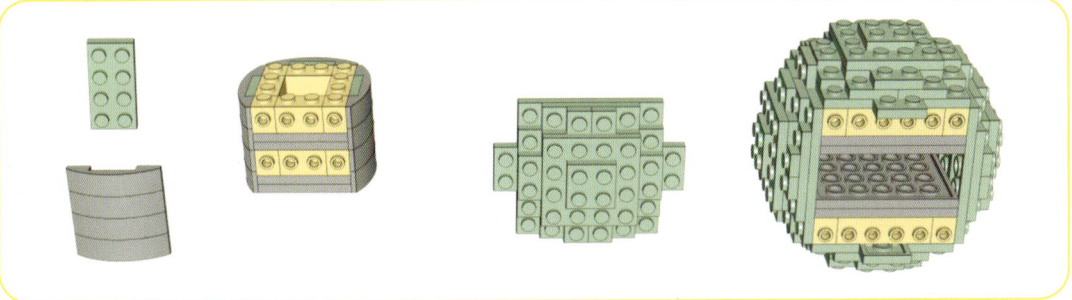

Figure 5-14: Cylinder and sphere shapes created by attaching curved slopes and panels composed of plates, respectively, to SNOT cores

It isn't possible to cover all the different ways that SNOT bricks can be used in LEGO builds, but hopefully these examples have given you a taste of what's possible and piqued your interest in exploring other applications for these elements.

PLATES WITH STUDS ON THEIR SIDES

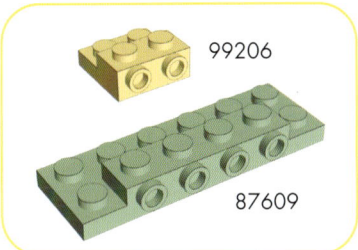

Figure 5-15: Plates with studs on their sides

Another type of SNOT element is the plate with studs on its side. These modified plates have a section that is 2 plates thick where the sideways studs are located (see Figure 5-15). They come in 2×2 (#99206) and 2×6 (#87609) versions, though other varieties may be added in the future.

To understand how these elements are useful, recall that you need to sandwich two layers of plates between each layer of SNOT bricks to properly space the sideways studs on the face of a wall. Because the SNOT portion of the modified plates is 2 plates thick, one of these elements can replace the two layers of regular plates, adding an extra, correctly spaced row of studs between each layer of bricks (see Figure 5-16). The resulting wall is similar to one built using the special SNOT bricks that are 5 plates high (#22885, for example), but with a key advantage: the modified plates are 2 studs deep.

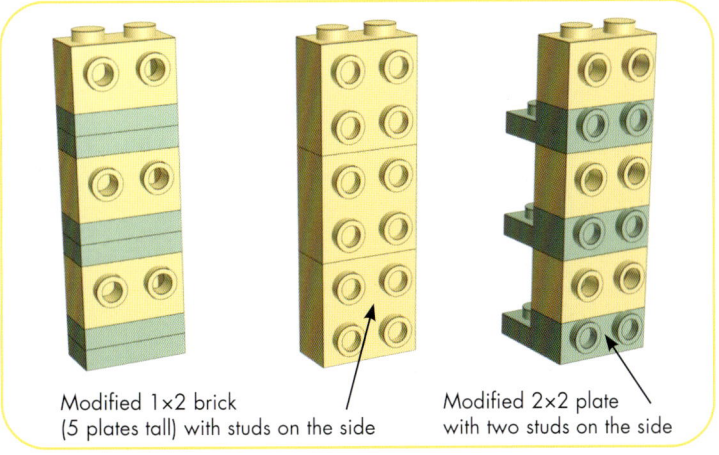

Modified 1×2 brick (5 plates tall) with studs on the side

Modified 2×2 plate with two studs on the side

Figure 5-16: Three different options for creating studs on the face of a wall

The added depth gives us the ability to use long plates in the back and strap the SNOT section to the bricks on either side for greater structural stability. Notice in Figure 5-17 how these elements give us a full complement of studs on the front of the wall, allowing us to attach 1×2 cheese slope pieces (#85984) for architectural detail. At the same time, the studs on the rear portion of these modified plates allow them to be tied securely into the rest of the wall.

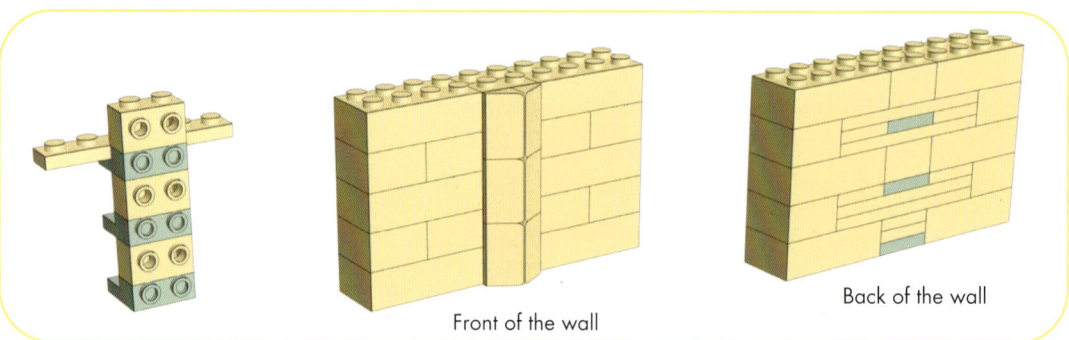

Front of the wall

Back of the wall

Figure 5-17: The front and back of a wall that uses plates with studs on their sides

The smaller SNOT plates (#99206) have been indispensable in some of my builds, including my Chrysler Building model. One of the most recognizable elements of the New York City skyline is the crown of the Chrysler Building. The iconic stainless-steel crown is made up of terraced arches that taper and culminate in a spire. Figure 5-18 shows how I used the smaller SNOT plates to create the curved panels that make up the building's crown.

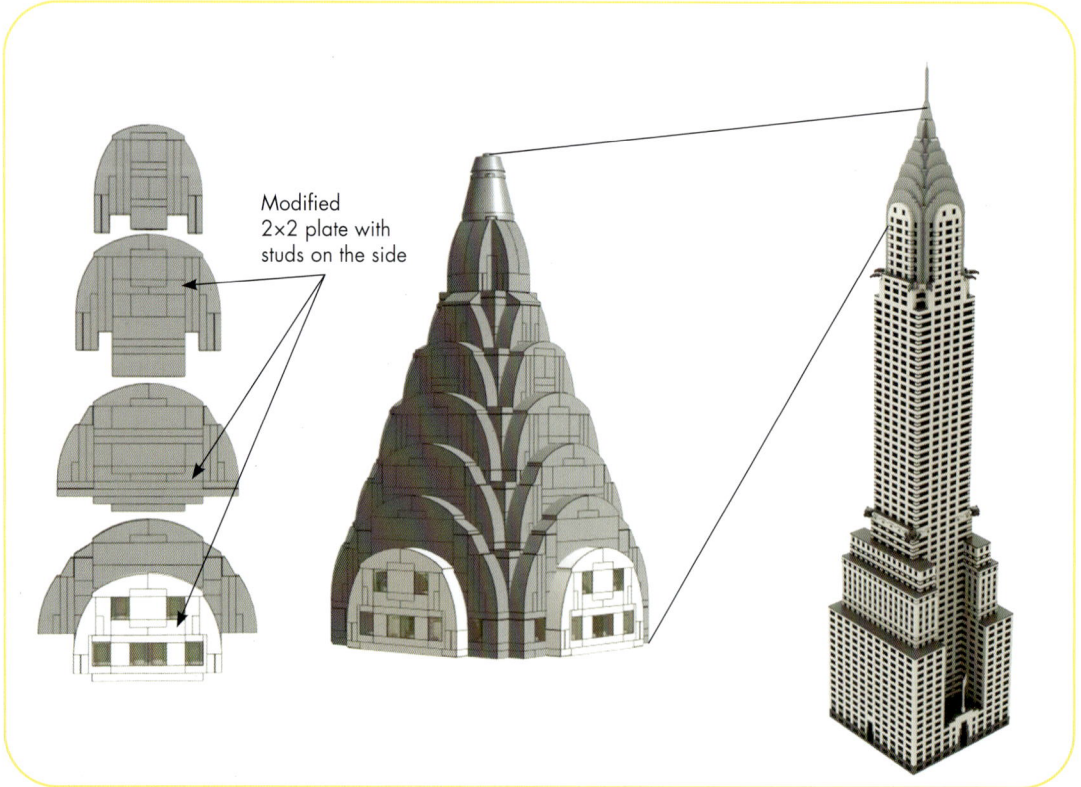

Modified 2×2 plate with studs on the side

Figure 5-18: Curved panels used to create the crown of the Chrysler Building

For each curved panel, I was able to combine SNOT elements (both bricks and plates) with regular bricks and plates in such a way that the overall assembly had studs on three sides in just the right locations. This allowed me to attach curved slope (and some cheese slope) pieces in both studs-up and sideways orientations, creating the best possible approximation of the cascading curves at the crown of the building.

BRACKETS

A LEGO *bracket* is essentially a 1×1 or 1×2 plate with studs on its side, except these studs are on an extension that's perpendicular to the plate. The extension has the same length and width as a normal plate, with anywhere from 1 to 8 studs (depending on the type of bracket), but it's only half as thick (see Figure 5-19).

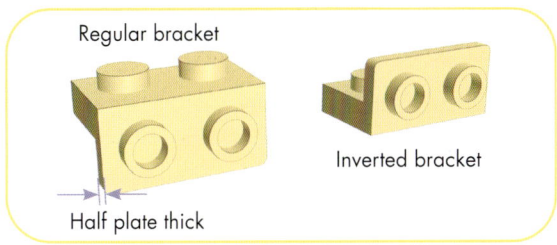

Figure 5-19: Simple brackets (regular and inverted)

There are two types of brackets: regular and inverted. In a *regular bracket*, the top edge of the extension is flush with the top of the plate (not including the studs), while in an *inverted bracket*, the bottom edge of the extension is flush with the bottom edge of the plate. Figure 5-20 shows a selection of some of the brackets available in the LEGO catalog.

Figure 5-20: A selection of brackets in the LEGO catalog

Brackets can be attached above or below structures built using regular bricks or plates, as shown in Figure 5-21. They make additional studs available in a sideways orientation, allowing us to add details to the structure or even sculpt its outer shape using slope and curved slope pieces.

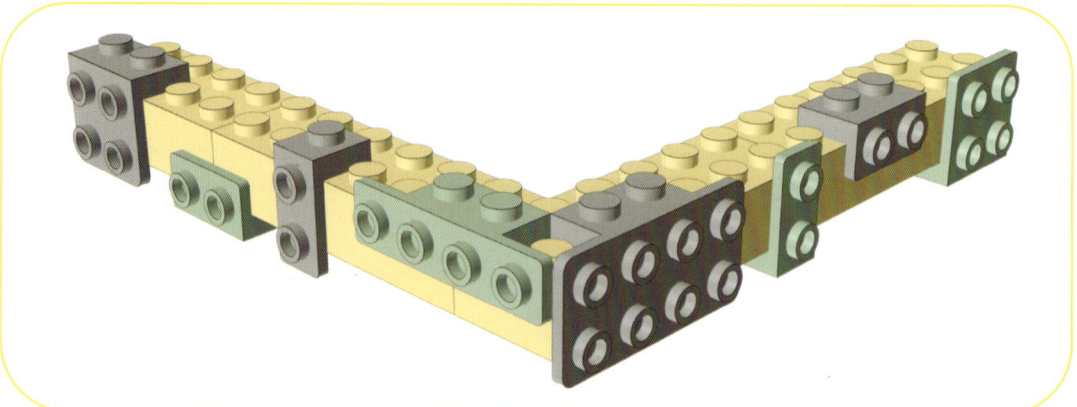

Figure 5-21: Regular and inverted brackets can be attached to the top and bottom, respectively, of other LEGO structures.

As we'll discuss next, the half-plate thickness of the sideways portion of a bracket can also be quite handy for creating half-plate offsets.

HALF-PLATE OFFSETS

The unique geometry of SNOT elements like headlight bricks and brackets makes it possible to offset or shift a LEGO element by half a plate relative to the predominant grid. This technique isn't to be confused with the half-stud offsets covered in Chapter 4; half a plate is a smaller amount than half a stud.

To understand why half-plate offsets may be needed, consider again that the LEGO system is based on a square grid where each square is 1 stud (or 2.5 plates) on a side. When we build sideways, however, the smallest increment normally available is 1 plate. This can sometimes lead to a half-plate gap or misalignment, especially if the horizontal portion spans an odd number of studs. You don't typically have to worry about misalignment when using SNOT to add detail or decoration to the exterior surfaces of a model. But if you're trying to seamlessly integrate a SNOT section into a model mostly built with studs on top, you need to pay close attention to how the sideways elements line up with the LEGO grid.

EVEN OR ODD STUD WIDTHS

As long as you can get the SNOT section to occupy an even number of studs, that works out to a whole number of plates (2 studs = 5 plates), so it'll be easy to avoid misalignment. The Taj Mahal in India is perhaps the most well-known example of Islamic architecture. It consists of a large marble mausoleum topped by a magnificent dome and accentuated by minarets. One of the modern wonders of the world, it features an array of stunning and ornate details. In my model of the Taj Mahal, I needed an accent stripe going around the main doorway to the mausoleum. I created the vertical portion of this stripe by attaching sand green plates sideways (see Figure 5-22). For this, I made a 2-stud-wide gap in the wall around the doorway and filled it with a brick and a tile placed sideways, in addition to the sand green plate used for the accent stripe itself. That way, the SNOT portion had a total thickness of 5 plates, exactly the same size as 2-stud gap in the wall.

Attached sideways

Figure 5-22: SNOT used to create an accent stripe around the doorway on the Taj Mahal

However, if the SNOT portion takes up an odd number of studs, it may end up being half a plate too wide or too narrow. Consider the example in Figure 5-23, which is loosely based on the roof section of my model of the Chrysler Building. Working upward from a 1×6 brick, we want to create a rounded taper on each side of a window by attaching curved slope pieces (#11477) sideways. For aesthetic reasons, the curve should flow smoothly from the straight portion built with studs on top. The window takes up 2 studs, and each SNOT brick takes up another stud, leaving just 1 stud on each side for the horizontal portion. The curved slope (combined with a 1×1 plate) is 2 plates thick, so if we attach it directly to the SNOT brick, the curved portion of the wall will be inset by half a plate relative to the straight portion, which is 2.5 plates (1 stud) thick (Figure 5-23, center).

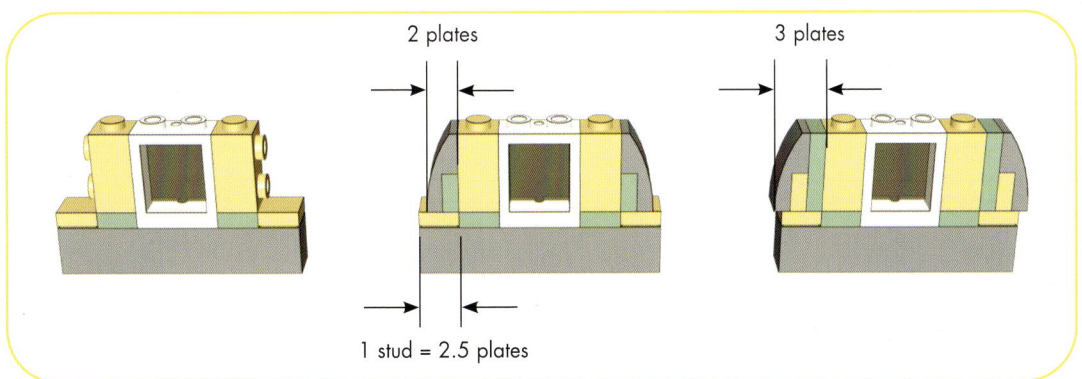

2 plates

3 plates

1 stud = 2.5 plates

Figure 5-23: A half-plate misalignment between the studs-up portion and the SNOT portion of the build

Adding a plate to the SNOT section doesn't help (Figure 5-23, right). Now the horizontal portion is 3 plates thick, so the curve sticks out by half a plate relative to the brick beneath it. What we need here is a half-plate offset: the equivalent of a brick with a stud on its side, but with the stud pushed out (or recessed) by half a plate. In fact, we can create just that using brackets or headlight bricks.

Recall that brackets have extensions that are half a plate thick. Stack a 1×1 bracket (#36841) on top of two 1×1 plates, and you get the equivalent of a 1×1 brick with a stud on its side, but the result is 3 plates thick instead of 2.5. A headlight brick, on the other hand, has a stud on its side, but the stud is recessed by half a plate, making it just 2 plates thick. Figure 5-24 shows both of these options alongside a regular 1×1 SNOT brick. Moving from left to right, notice that each horizontal stud is offset by half a plate compared to its neighbor.

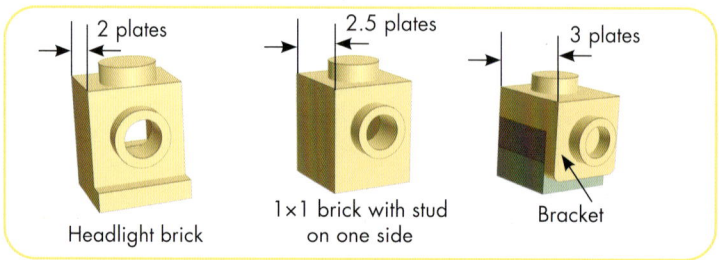

Figure 5-24: Three different options with half-plate incremental offsets in the stud location

With this in mind, let's return to our tapered wall example. By using inverted brackets instead of SNOT bricks, we can push out the horizontal section by half a plate. That way, the curved slopes will line up correctly with the straight wall below them (see Figure 5-25).

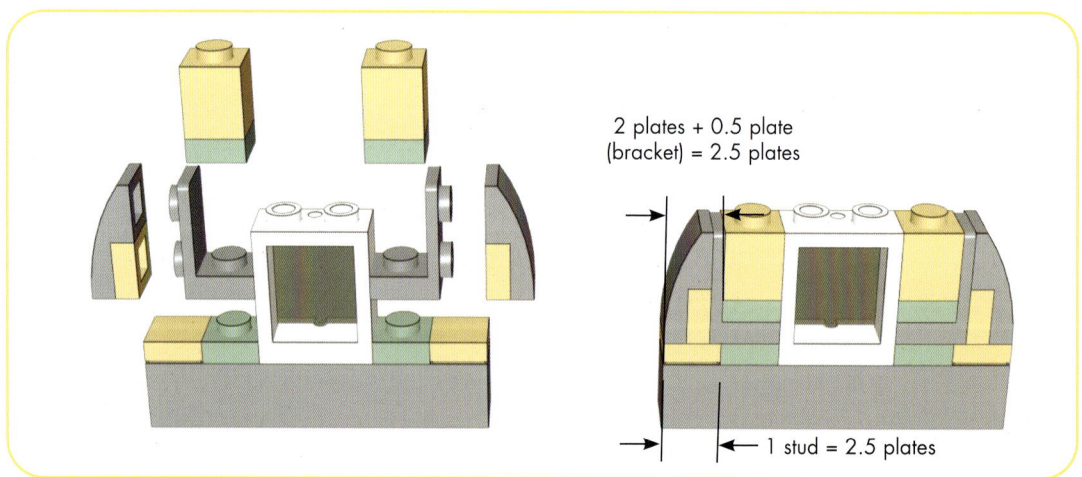

Figure 5-25: Using brackets to get proper alignment between the studs-up and SNOT portions of the build

CASCADED CHEESE SLOPES

Half-plate offsets can also eliminate the jaggedness in slopes built by cascading multiple cheese slope pieces. As we've seen, these 1×1 slopes are 2 plates high, but they have a half-plate-high lip at the base of the slope. If we were to simply stagger each successive cheese slope by 2 plates, the resulting slope wouldn't be smooth because of the stair-stepping effect from the half-plate lip (see Figure 5-26, left). What we need here is a way to stagger each cheese slope by 2 − 0.5 = 1.5 plates (see Figure 5-26, right).

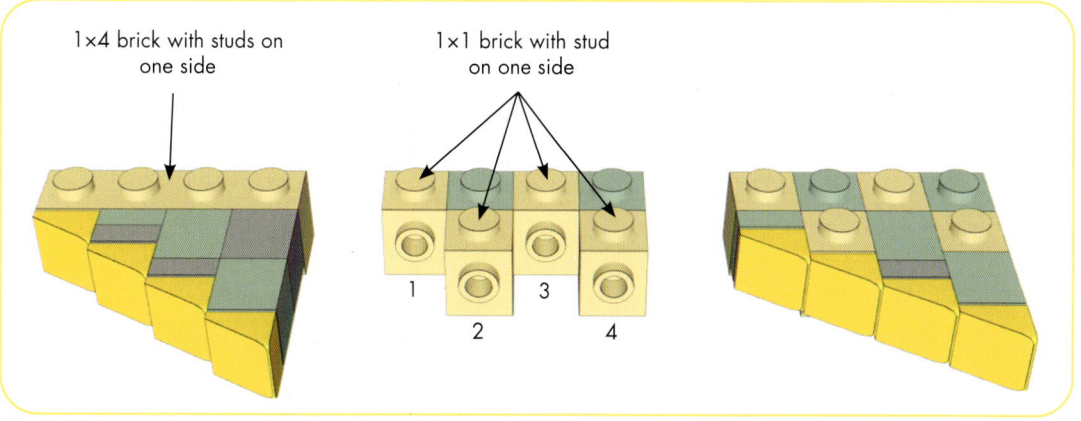

Figure 5-26: Using the geometry of bricks and plates to create a 1.5-plate offset

Here's how it works. A regular brick is 2.5 plates deep, and 1 plate less than that is 1.5 plates. We can place a sequence of 1×1 bricks with a stud on their sides, as shown in the middle of Figure 5-26. The even-numbered bricks (2, 4) are pushed forward by 1 stud (2.5 plates) by placing a regular 1×1 brick behind them. Next, attaching a sideways plate to the front of brick 1 gives us a stud that's sticking out 2.5 – 1 = 1.5 plates less than the stud on brick 2—exactly the offset we need. For bricks 3 and 4, we add a sideways 1×1 brick for an extra 3 plates of depth, plus another sideways plate for brick 3, again for a 1.5-plate offset. Counting the total number of plates of offset we have in the sequence, we get:

1. 1 plate
2. 2.5 plates (1 stud)
3. 4 plates (1 brick + 1 plate)
4. 5.5 plates (1 stud + 1 brick)

Each of these is separated by 1.5 plates compared to the previous one in the sequence.

We can also use the half-plate offsets provided by headlight bricks and brackets to achieve the same effect, as in Figure 5-27.

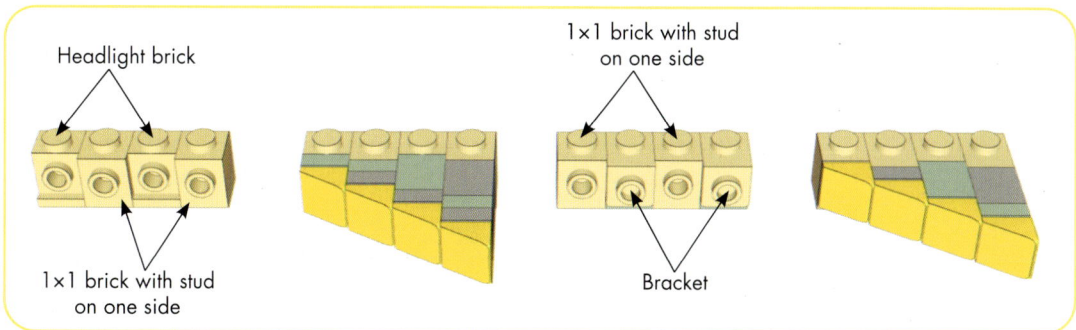

Figure 5-27: Creating the 1.5-plate offsets using headlight bricks (left) and brackets (right)

On the left side of the figure, you see headlight bricks alternated with 1×1 bricks with a stud on one side. Given that the stud on the face of the headlight brick is recessed by half a plate, all we need to do is attach an extra plate to the brick with the stud on its side to get a 1.5-plate offset relative to the headlight brick. On the right, the same offset is achieved using brackets that have a stud sticking out by half a plate relative to a regular SNOT brick. Again, we add an extra plate to the stud on the bracket to get the 1.5-plate offset.

QUARTER-PLATE OFFSETS

It's possible, though perhaps not very practical, to combine SNOT techniques with the half-stud offsets created using jumper plates (see Chapter 4) to get a quarter-plate offset. Since 1 stud is the same as 2.5 plates, the half-stud offset of a jumper plate is equivalent to 2.5/2 = 1.25 plates.

Start with two identical SNOT bricks—say, headlight bricks—placed next to each other. Then offset one by half a stud using a jumper plate, and attach a 1×1 plate to the front of the other. The studs in the front end up 1.25 – 1 = 0.25 plates apart, as shown in Figure 5-28.

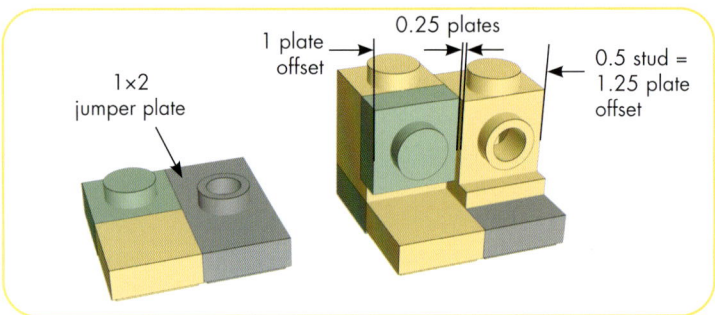

Figure 5-28: Using a jumper plate to create a quarter-plate offset

Combining the half-plate offsets from headlight bricks and brackets with the quarter-plate offsets from jumper plates, we can create longer sequences of horizontal studs, each offset by an extra quarter plate compared to the one before, as in Figure 5-29.

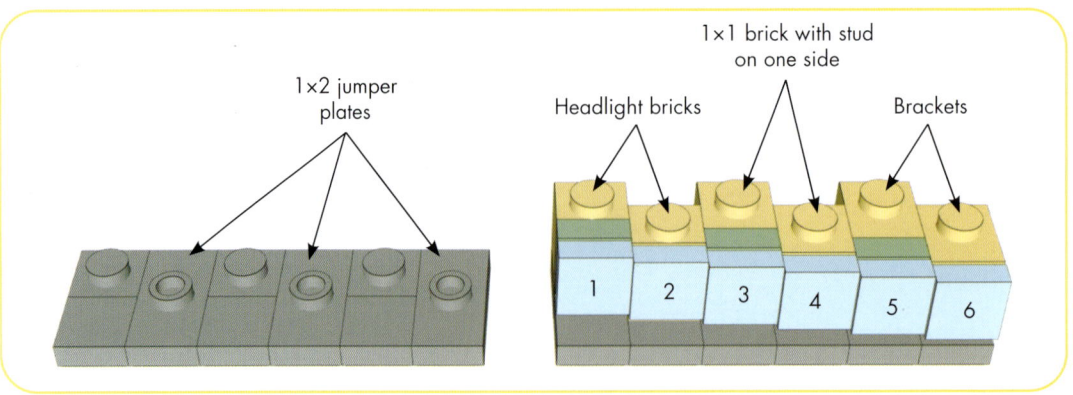

Figure 5-29: A sequence with quarter-plate offsets on each successive step

Here's how the math works for this sequence:

1. Headlight brick (2 plates) + 1 plate = 3 plates
2. Half-stud offset using jumper plate (1.25 plates) + headlight brick (2 plates) = 3.25 plates
3. Brick with stud on side (2.5 plates) + 1 plate = 3.5 plates
4. Half-stud offset using jumper plate (1.25 plates) + brick with stud on side (2.5 plates) = 3.75 plates
5. Bracket (3 plates) + 1 plate = 4 plates
6. Half-stud offset using jumper plate (1.25 plates) + bracket (3 plates) = 4.25 plates

I have yet to find many applications for this technique, at least in my models, but perhaps you'll be able to find interesting uses for quarter-plate offsets.

STUD REVERSAL

When we build sideways, we typically turn the direction of the studs by 90 degrees relative to their normal orientation, but there may be situations where we instead need to turn the direction of the studs by 180 degrees, essentially reversing their direction. Another classic skyscraper in Lower Manhattan is 40 Wall Street with its distinctive, Gothic-inspired green pyramidal roof (see Figure 5-30). My model of this building features four panels (composed of plates) that are angled using hinge plates (which you will learn more about in Chapter 6). Each panel is built in two halves with studs facing opposite directions. Viewed from the front (Figure 5-30, center), all you see are the vertical lines from the rows of plates turned on their sides. Viewed from the back (Figure 5-30, right), you can see how the two halves of the panel are joined together using bricks with studs on one side.

Figure 5-30: A roof panel used on the model of 40 Wall Street

There are many ways—some legal, others not—to reverse studs. Figure 5-31 shows a few legal techniques using the SNOT elements we've covered in this chapter.

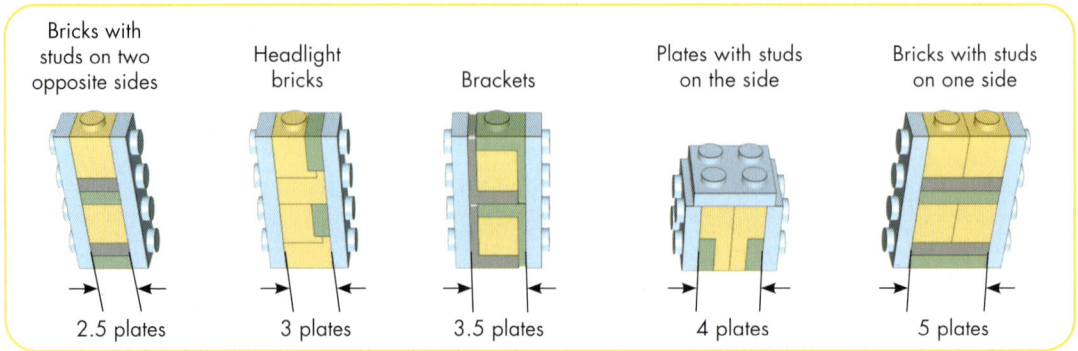

Figure 5-31: Different legal options to reverse studs

The technique you choose depends on the spacing you need between the reversed studs. Using bricks with studs on two opposite sides (#47905), you can achieve a width of 2.5 plates (1 stud). Combining sideways and upright headlight bricks gives you a width of 3 plates, and mixing regular and inverted brackets gives you a width of 3.5 plates. Two back-to-back plates with studs on the side yield a width of 4 plates, but note that the bottom of this assembly doesn't have any attachment points. Finally, back-to-back bricks with studs on one side give a width of 5 plates, or 2 studs.

SNOT WITH HALF-STUD OFFSETS

The sideways studs on the SNOT elements we've reviewed so far all line up with the elements' vertical studs, but with a little creativity, you can make your own SNOT elements featuring half-stud offsets. Simply take a 1×2 Technic brick with 1 hole (#3700) and insert a Technic half-pin (#4274) into the hole. This creates a stud in the middle of the face of the Technic brick, between the two vertical studs, similar to having a sideways jumper plate (see Figure 5-32).

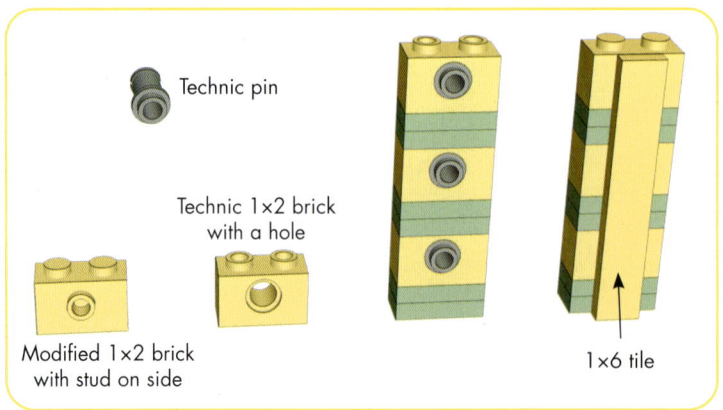

Figure 5-32: Options for attaching elements sideways with a half-stud offset

Attaching tiles and other elements to these offset SNOT studs can add subtle detail to a build. For example, I used this technique in the top section of my Empire State Building model (Figure 5-33) to continue the vertical lines of columns from the section below, even as the walls themselves are offset half a stud from the section below.

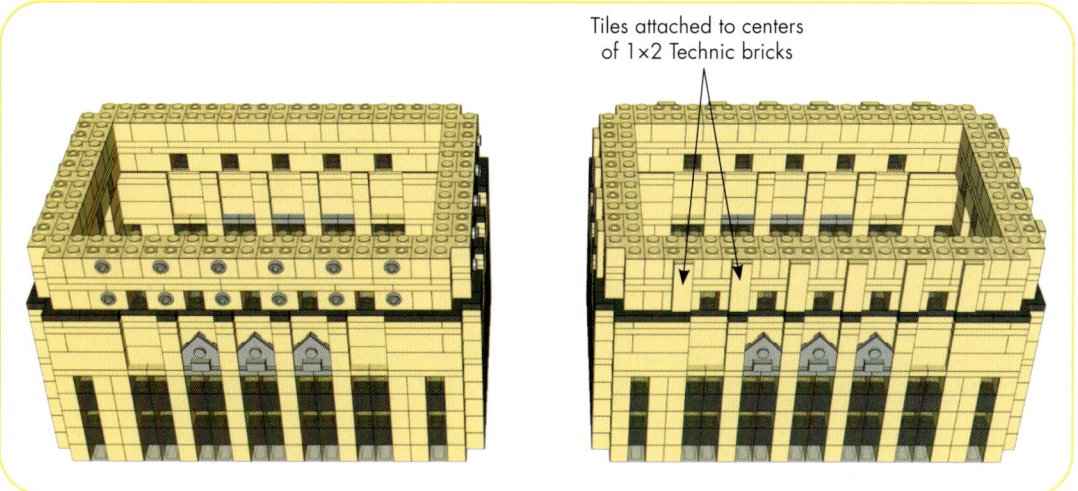

Figure 5-33: SNOT with half-stud offsets used on the top section of the Empire State Building

In 2022, LEGO came out with a 1×2 brick with a stud on the middle of its side (#86876), equivalent to the Technic brick-and-pin combination. If these bricks aren't yet available in the color you need, you can make your own using Technic bricks.

ILLEGAL SNOT TECHNIQUES

We've focused on legal SNOT techniques in this chapter, but there are some illegal techniques as well. I alluded to one in Chapter 3, where I showed how you can insert a stud (albeit with a very tight fit) into the hole on the side of a Technic brick, and I mentioned this technique would be useful in the Empire State Building model we were working on.

When we left off with this model in Chapter 3, we had stacked a few floors to create a small portion of the largest section of the building, using layers with alternating crosswise and lengthwise orientations. After some effort, we managed to arrive at a fully interlocking structure. Well, almost. As shown in Figure 5-34, each corner of the model has a tall stack of tan 1×1 bricks, and because the bricks on either side of them have other colors, there's no easy, legal way to tie these stacks into the main structure of the building.

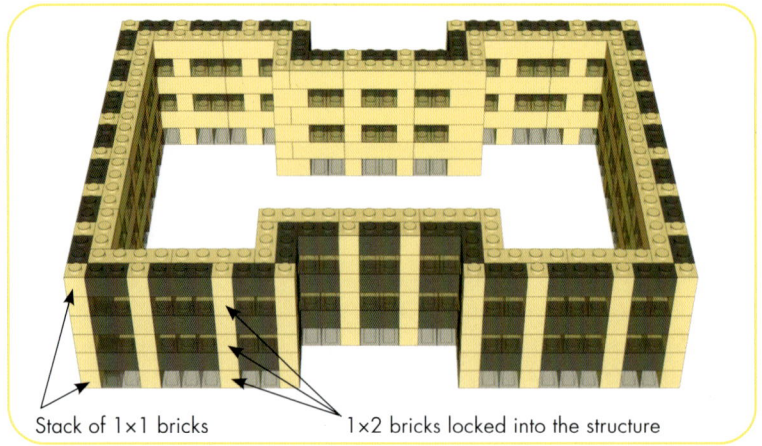

Stack of 1×1 bricks   1×2 bricks locked into the structure

Figure 5-34: Three floors of the biggest section of the Empire State Building, with loose stacks in the corners

Here's where the illegal SNOT technique comes to the rescue. We can start with a 1×1 brick with studs on two adjacent sides and attach a 1×1 Technic brick with a hole to each of those two sideways studs (see Figure 5-35). This creates a "corner brick" that combines two colors (tan for the corner and dark bluish gray for the window accents).

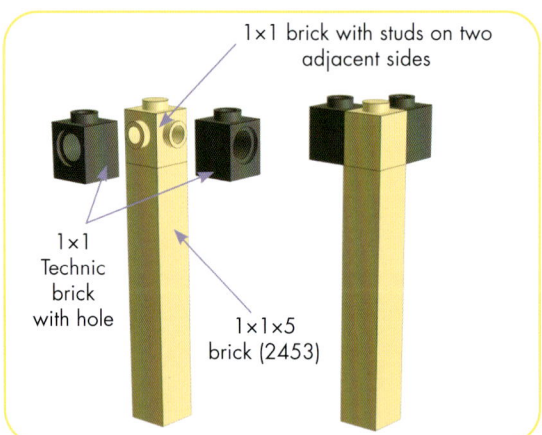

1×1 brick with studs on two adjacent sides

1×1 Technic brick with hole

1×1×5 brick (2453)

Figure 5-35: A corner brick with two colors created using a SNOT brick and two Technic 1×1 bricks with holes

By putting this corner brick assembly on top of a tall 1×1×5 brick, we only need to resort to the illegal technique every sixth layer of bricks. Figure 5-36 shows the assembly integrated into the corners of the model. The corner bricks on the next section below will lock the 1×1×5 bricks into place.

The problem with this technique is that the center point of the sideways stud on a 1×1 SNOT brick isn't perfectly aligned with the center point of the hole on a 1×1 Technic brick. The former is 3.92 mm from the top of the brick, while the latter is 3.8 mm from the top of the brick. This misalignment can stress the elements used, especially if we stack other bricks or plates on top of this combination. In the case of the Empire State Building, however, we have no other choice if we want to create a fully interlocking structure.

Figure 5-36: Using the corner brick to avoid having tall stacks of 1×1 bricks in the corners

SUMMARY

This chapter introduced sideways building using SNOT (studs not on top). We discussed how the 6:5 height-to-width ratio of LEGO bricks influences sideways building, explored a variety of SNOT elements, and examined some SNOT-specific techniques such as creating half- and quarter-plate offsets. But this is just the tip of the SNOT iceberg. Once you start building horizontally as well as vertically, there's a whole world of techniques to explore, especially if you expand beyond basic system SNOT elements like bricks, plates, and brackets and start incorporating Technic and other specialized parts. The SNOT fundamentals in this chapter can serve as a jumping-off point for your exploration of other sideways techniques. In the next chapter, we'll turn our attention from sideways building to angled walls.

6

ANGLED WALLS

We've seen that the studs on a LEGO baseplate are laid out in a regular square grid, which seemingly limits the way pieces can be attached. If you stick to this grid, the walls of your LEGO structure can only form 90-degree angles with each other. But suppose you want to build a wall on a diagonal. Are you out of luck? No! In this chapter, we'll explore how to break free from the LEGO grid by building angled walls. We'll cover LEGO elements that can be used for this, as well as tips for building angled walls digitally in BrickLink Studio.

THE LEGO GRID GEOMETRY PROBLEM

To understand why building angled walls presents a challenge in the first place, it helps to think about the geometry of the LEGO grid. Consider a square made up of four studs on a baseplate, as in Figure 6-1. The distance between the studs at any two adjacent corners will be exactly the same. It's 0.8 cm, or 1 stud, our favorite basic unit of LEGO measurement. However, the distance between studs at opposite corners of the square is greater. It's $\sqrt{2} \times 0.8$ cm = 1.414×0.8 cm = 1.13 cm, or 1.414 studs. That's the nature of squares: the distance between any two opposite corners is a little longer than the length of each side. As a result, although it's possible to attach a 1×2 LEGO brick on top of two adjacent studs in the square, it isn't possible to attach that brick on top of two diagonal studs—the studs on the baseplate and the anti-studs on the underside of the brick won't line up.

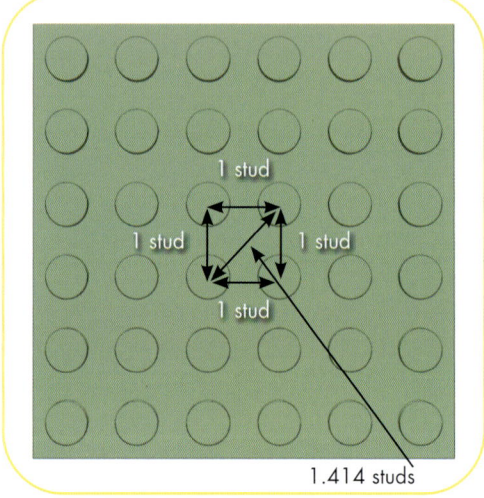

Figure 6-1: The geometry of the LEGO grid

Similarly, the distance between any two studs measured at any angle other than 0 or 90 degrees isn't guaranteed to be a whole number of studs. Figure 6-2 shows some examples of this. As long as the distance between two studs on a diagonal isn't a whole number of studs, you won't be able to connect them with a LEGO brick.

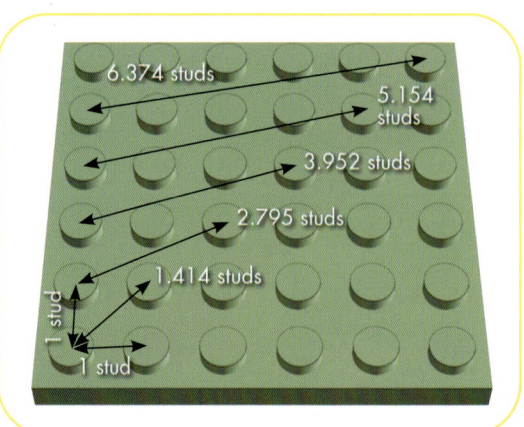

Figure 6-2: The distances between studs measured at different angles

THE PYTHAGOREAN SOLUTION

How can we create an angled wall that attaches firmly to a baseplate when the math is stacked against us? Let's try a little experiment to find out. Place two 1×1 plates on a base-plate, as shown in Figure 6-3. Then try placing a 1×6 brick diagonally to bridge these two 1×1 plates.

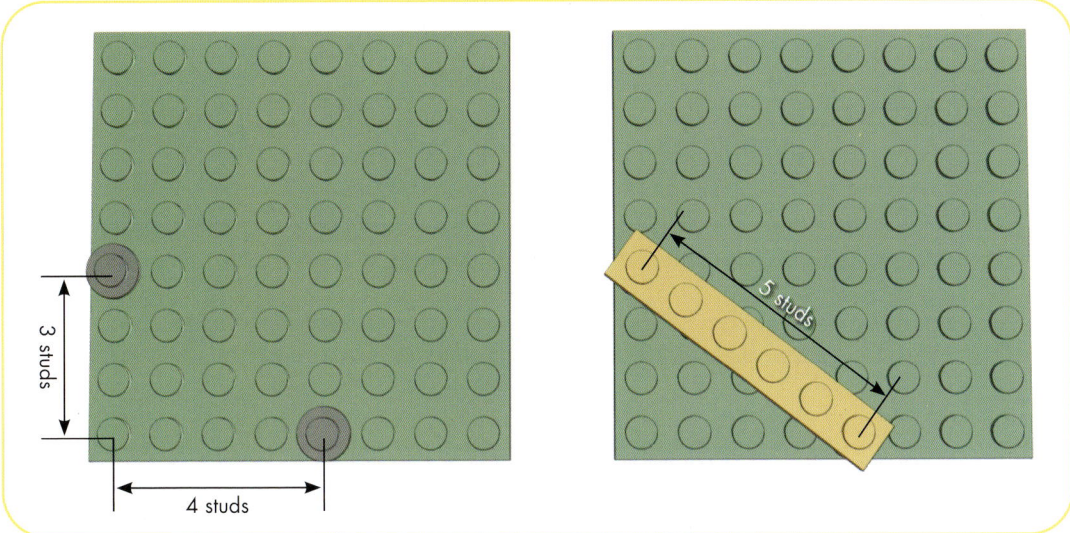

Figure 6-3: Attaching a 1×6 brick at an angle

It works! The studs at the two ends of the 1×6 brick line up with the studs on the two 1×1 plates. This allows the 1×6 brick to have a good connection to the baseplate (at least at the two ends). The remaining four studs on the 1×6 brick don't line up with the studs below. This is why we had to use the 1×1 plates as spacers to raise the 1×6 brick just enough to clear the remaining studs of the baseplate underneath.

What exactly is going on here? If you jog your memory back to high school math (if you haven't gotten there yet, you'll just have to take my word for it), the equation $a^2 + b^2 = c^2$ may sound familiar. This is the *Pythagorean theorem*. It defines the relationship between the sides of a right triangle, as shown in Figure 6-4. Here a and b are the lengths of the two sides that form a right angle, and c is the length of the third, diagonal side, known as the *hypotenuse*. The hypotenuse is always the longest side of a right triangle.

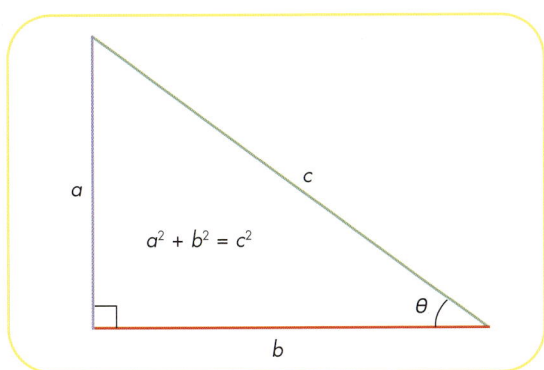

Figure 6-4: The Pythagorean theorem illustrated

Take a closer look at where the two 1×1 plates are placed on the baseplate in Figure 6-3. If you count along the two sides of the baseplate starting from the corner, you'll find that they're 3 and 4 studs away from the corner stud, and that they make up sides *a* and *b* of a right triangle. The 1×6 brick forms the hypotenuse of that triangle, and the distance between the studs at the two ends of the brick is 5 studs. Doing the math, we get $3^2 + 4^2 = 5^2$ (or $9 + 16 = 25$). We've created a right triangle that satisfies the Pythagorean theorem.

PYTHAGOREAN TRIPLES

The bottom line is that for any brick or plate to be placed at an angle other than 0 or 90 degrees, you need to make sure the resulting right triangle satisfies the Pythagorean theorem and that each side of the triangle is a whole number of studs long. Any set of three whole numbers that satisfies the Pythagorean theorem is called a *Pythagorean triple*. The one we've used, (3,4,5), is the smallest possible Pythagorean triple, but there are others. Table 6-1 lists all the Pythagorean triples with side lengths less than or equal to 25.

TABLE 6-1: Pythagorean Triples

Side *a*	Side *b*	Hypotenuse	Angle (degrees)
3	4	5	36.8
5	12	13	22.6
6	8	10	36.8
7	24	25	16.2
8	15	17	28.0
9	12	15	36.8
12	16	20	36.8

The last column in the table shows the smallest angle in the resulting right triangle. (For the other angle in the triangle, subtract the angle shown in the table from 90.) It's sometimes useful to know these angles—for example, if there's a particular angle you want to achieve, or if you're building angled walls digitally (more on this later in this chapter). To determine the angles in a right triangle, we need to use a little basic trigonometry.

Look again at the right triangle in Figure 6-4 and consider the angle labeled θ (the Greek letter theta). The ratio of the length of side *a* (the side opposite this angle) to the length of side *b* (the side adjacent to this angle) is called the *tangent*, and this ratio is fixed for a given angle regardless of the size of the triangle. If we know the side lengths but not the angle, we can figure out the angle using the inverse of the tangent function, known as the *arctangent* or *arctan* (available in most scientific calculators). For a (3,4,5) triangle, for example, the tangent of the smallest angle is $3/4 = 0.75$, and the arctan of 0.75 is 36.8 degrees.

Not all of the triples listed in Table 6-1 have practical applications in LEGO builds; in some cases, the numbers involved are too big, or the angles they help create are too narrow to be

Figure 6-5: The LEGO Boutique Hotel set (10297)

useful. However, basic triples like (3,4,5) are a popular way to create angled walls. For example, LEGO's official Boutique Hotel set (10297), shown in Figure 6-5, uses the (3,4,5) triple extensively.

Figure 6-6 shows that the set uses not one but six separate (3,4,5) triangles (including two that intersect each other) to create the angled walls that give the building its unique triangular shape.

Given that the hypotenuses of triangles 1 through 4 are in a straight line, connected using long plates, we don't even need to attach the outer ends of triangles 1 and 4 to the base. In particular, the angled wall can stop partway along the hypotenuse of triangle 4, at the point where it lines up with the hypotenuse from triangle 6, which forms the doorway to the hotel. This works nicely because triangle 6's hypotenuse is at a right angle to the main hypotenuse of triangles 1 through 4. (The same goes for triangle 5.)

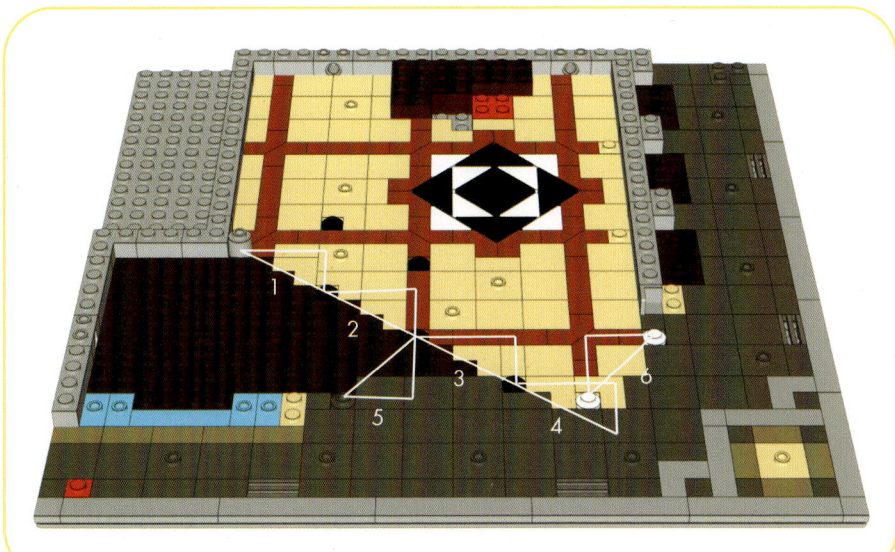

Figure 6-6: (3,4,5) triangles used for attaching the angled wall

USEFUL PIECES FOR ANGLED WALLS

LEGO elements with sections that swivel or rotate freely can come in handy for building angled walls. In this section, we'll look at two such elements—hinge plates and turntables—and see how they can be used.

HINGE PLATES

When you create an angled wall by placing a brick or plate diagonally on top of standard LEGO plates, there's no good way to avoid gaps at the corners where the angled wall segments meet the wall segments aligned with the LEGO grid, as shown on the left of Figure 6-7. An alternative is to use *hinge elements*, as in the right half of Figure 6-7. These pieces close the gaps completely on the inside of the wall and leave a much smaller gap on the outside.

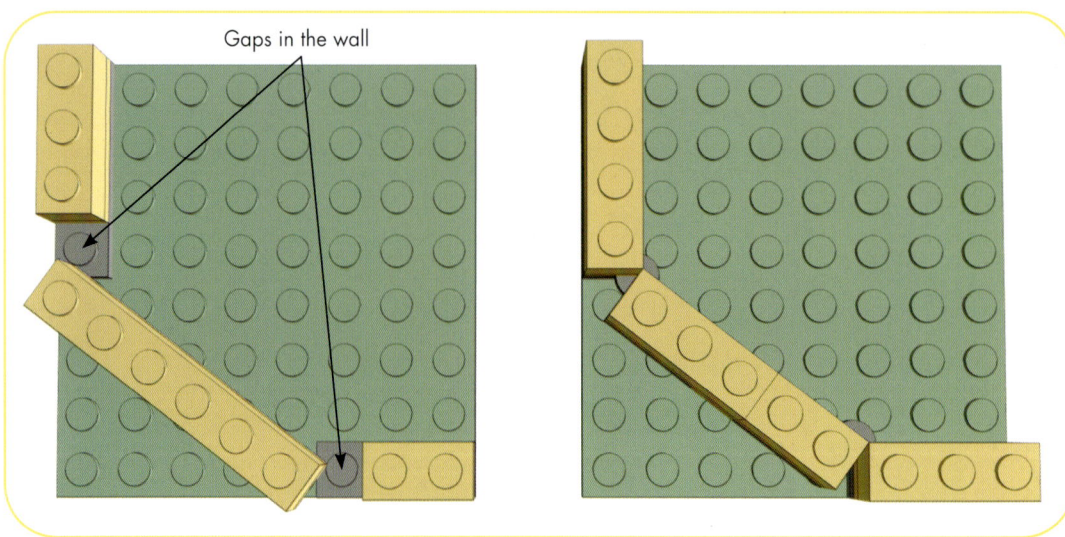

Figure 6-7: An angled wall created without (left) and with (right) hinge elements

There are many different types of LEGO hinge elements, but the ones you need for angled walls are the ones that swivel—specifically, the 1×4 hinge plate (see Figure 6-8). It consists of a 1×2 swivel base (#2429) and a 1×2 swivel top (#2430).

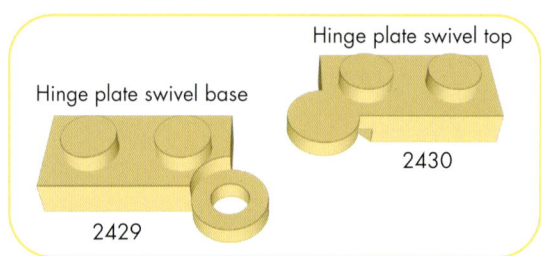

Figure 6-8: The parts of a hinge swivel plate

The two 1×2 plates that make up this element are joined at their corners by a hinge, allowing you to change the angle between the plates from 0 to 180 degrees, as seen in Figure 6-9.

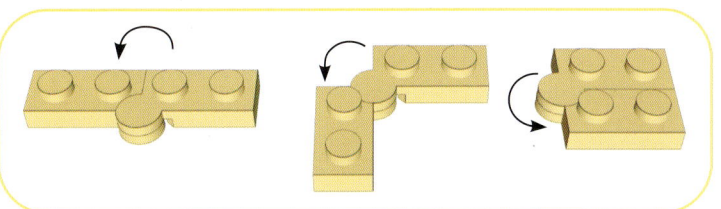

Figure 6-9: A hinge swivel plate can create angles between 0° (left) and 180° (right).

If you place two 1×4 hinge plates on a baseplate, as shown in Figure 6-10, you can bridge them with a 1×5 plate attached on top.

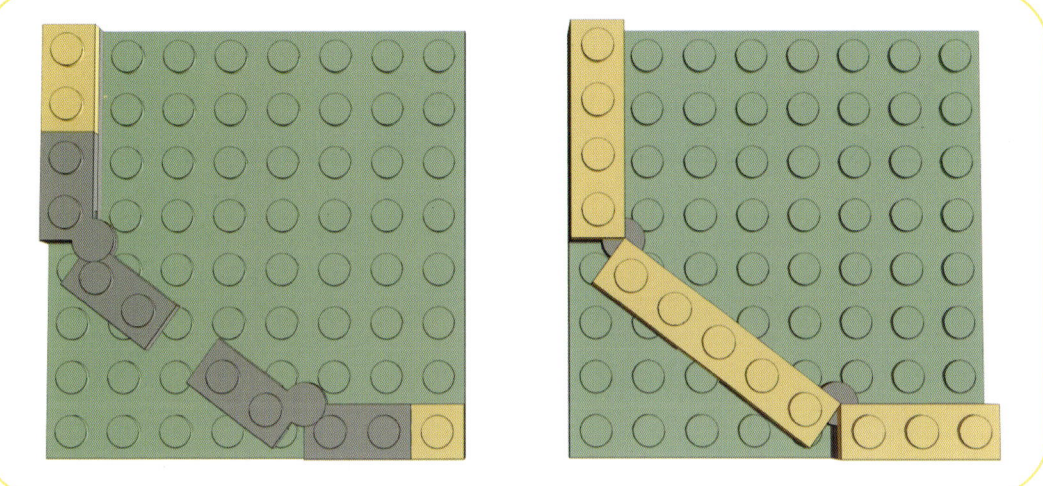

Figure 6-10: Creating an angled wall using hinge plates

We're still creating the same (3,4,5) right triangle as we were in Figure 6-3, but this time the hypotenuse plate has 5 studs instead of 6. This is because the sides of the triangle now intersect at the corners of the plates rather than the studs. Figure 6-11 illustrates the difference.

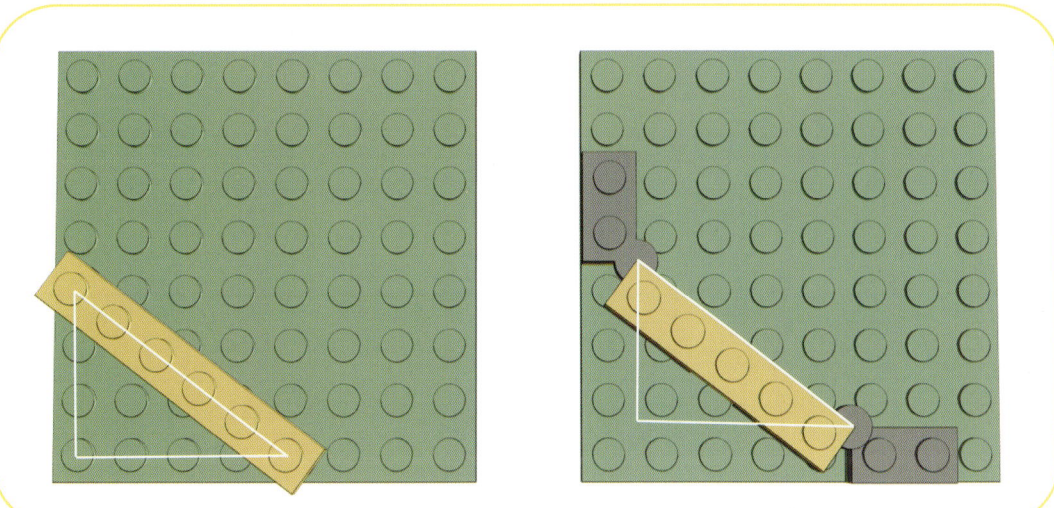

Figure 6-11: Two methods for creating angled walls

As you can see, the triangles in both cases are the same, with a hypotenuse that's 5 studs long. In the first case, this is measured between the studs at the ends of a 1×6 plate, while in the second case, it's measured between the corners of a 1×5 plate.

TURNTABLES

Turntables give us another way to create angled walls. A LEGO turntable consists of a base (2×2 or 4×4) that can be attached like a normal plate and a top that can swivel freely a full 360 degrees around (see Figure 6-12). The 2×2 base (#3680) requires a matching top element (#3679), while the 4×4 base (#61485) can accommodate a variety of compatible elements, including a 4×4 round plate (#60474).

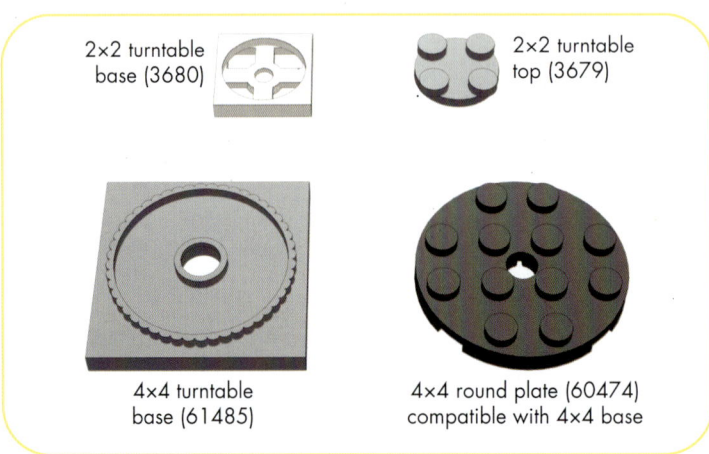

Figure 6-12: A selection of turntable pieces

We can form the same (3,4,5) triangle using 2×2 turntables. With the 2×2 bases attached to a baseplate, rotate the top elements such that a 1×5 plate can be connected across them, as shown in Figure 6-13. The sides of the (3,4,5) triangle now intersect at the axes of rotation on the turntables (the center points of the top plates).

Figure 6-13: A (3,4,5) triangle created using turntables

Multiple official LEGO sets harness turntables to build walls and other structures at an angle. An example is the Spring Lantern Festival set (80107), where 4×4 turntables are used to attach an arched footbridge at an angle over a koi pond (see Figure 6-14).

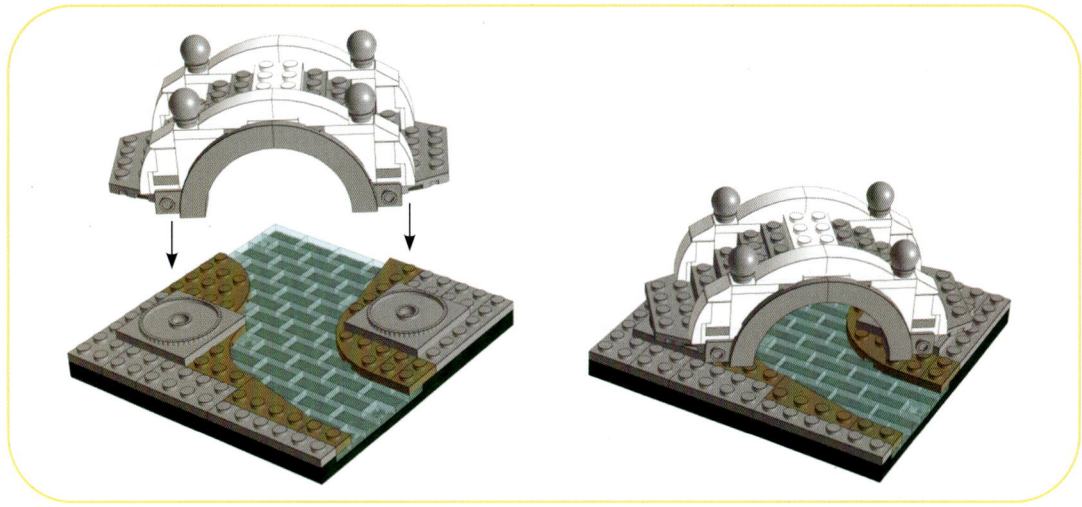

Figure 6-14: The LEGO Spring Lantern Festival set (80107)

Figure 6-15 shows a simplified version of the bridge from this set. The 4×4 turntable bases are attached on either side of the koi pond, and the top elements (4×4 round plates in this case) are rotated to be able to attach the bridge at an angle.

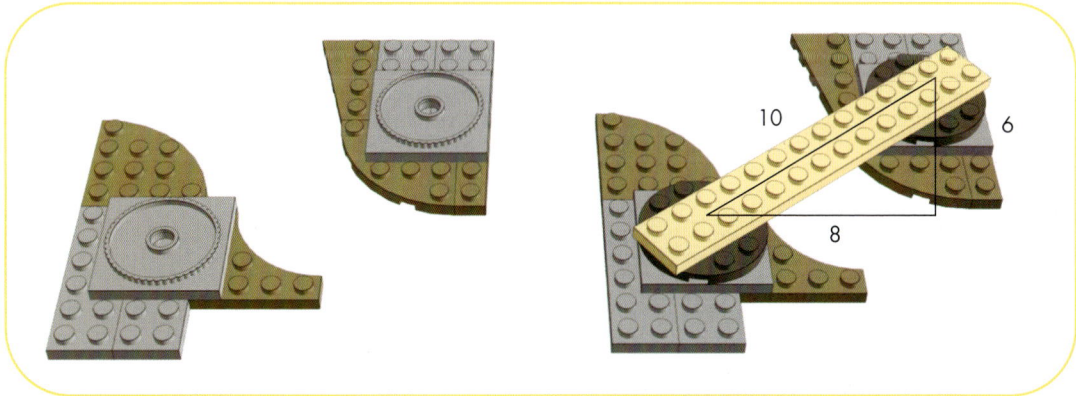

Figure 6-15: A bridge attached at an angle in the Spring Lantern Festival set

Measuring from the center points of the two turntables, the Pythagorean triple at work here is (6,8,10). Notice how this is simply a multiple of the (3,4,5) triple.

DIGITAL BUILDING TIP

To build an angled wall digitally in Studio, you'll need to rotate pieces more precisely than is possible with the arrow keys on your keyboard (which you can use only for rotations in 90-degree increments). To do this, select the piece(s) to be rotated; arrows will appear on either side of a center dot for each possible axis of rotation. Click one of the arrows and drag your mouse in the direction the arrow is pointing to rotate the piece that way. But this can sometimes get fiddly. Clicking one of the center dots will bring up a rotation circle that allows you to rotate with greater precision, as well as a text field where you can enter the exact angle of rotation. You can get this angle of rotation by looking it up in Table 6-1. For instance, the angle you need to enter to achieve a (3,4,5) triangle is 36.8 degrees. This same technique works for rotating the top plate of a turntable. If you are working with a hinge or turntable assembly where it is possible to rotate just one of the elements that make up the assembly, the hinge connection is detected automatically, and clicking the assembly will bring up a hinge icon. Click this icon to select the piece to be rotated relative to the other pieces in the assembly.

NEAR TRIPLES

We've seen how to take advantage of Pythagorean triples to build angled walls, but there are only so many of these triples to choose from. Thankfully, hinge plates have some give, which makes it possible to fudge the math a little bit and use numbers that are close enough to, but not quite, Pythagorean triples. These *near triples* open up more possibilities for angled building, allowing other combinations of side lengths and, perhaps more important, other angles.

45-DEGREE ANGLES

When a right triangle contains a 45-degree angle, it's called a *special right triangle*. As shown in Figure 6-16, the third angle also ends up being 45 degrees (the three angles in a triangle have to add up to 180 degrees), and the sides *a* and *b* that form the right angle have to be of equal length.

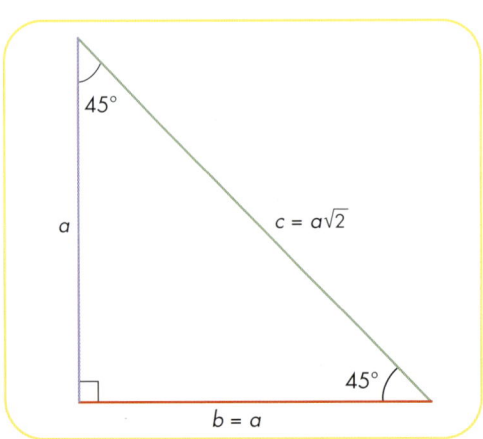

Figure 6-16: A special right triangle

There are no exact Pythagorean triples that form special right triangles, since the hypotenuse of such a triangle is always a multiple of the square root of 2 and will never be a whole number. In a pinch, you might be able to pass off the wall created by a (3,4,5) triangle as a 45-degree angle, but as we've seen, the small angle of that triangle is closer to 37 degrees, so the effect is far from precise. Here's where near triples come to the rescue. The near triples (5,5,7) and (7,7,10), shown in Figure 6-17, both produce 45-degree angles.

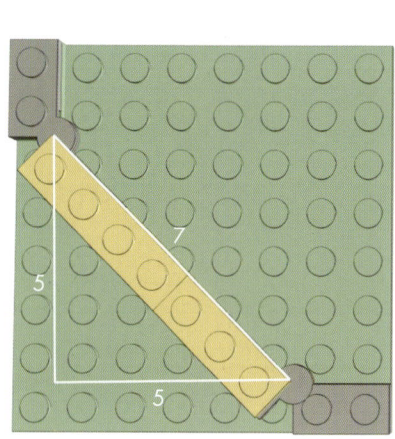 

Figure 6-17: Angled walls created using two near triples

Figure 6-18: The LEGO Corner Garage set (10264)

Consider the math for a moment: $5^2 + 5^2 = 50$, which is close enough to $7^2 = 49$ to work with the wiggle room that the hinge plate provides. Likewise, $7^2 + 7^2 = 98$, which is close enough to $10^2 = 100$. Other near triples for creating 45-degree angles include (12,12,17) and (17,17,24).

LEGO's official Corner Garage set (10264) uses the (12,12,17) near triple to create the part of the main building facade that sits at a 45-degree angle. The model is shown in Figure 6-18.

Figure 6-19 shows how the angled wall section is attached to the base. It's built on a 2×16 plate that's attached at a 45-degree angle using 1×2 round plates (#35480).

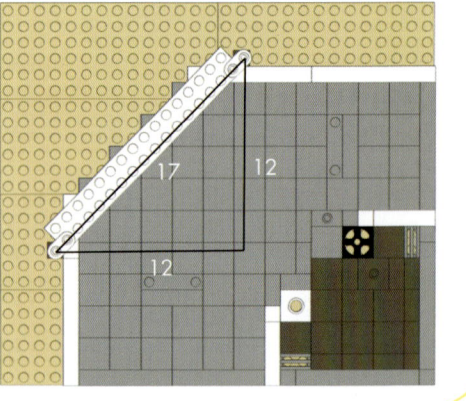

Figure 6-19: An angled wall (using a near triple) in the Corner Garage set

The total length of the angled section is 17 studs, measured between the studs at the two connection points. If we think of this as the hypotenuse of a right triangle, the other two sides would each be 12 studs long (if you picture horizontal and vertical lines drawn along the LEGO grid from the studs at the two connection points, they will intersect at a stud that is 12 studs away in each direction). The round plates act like hinges and provide a firm connection while allowing a little bit of wiggle room for this near triple.

I used a different near triple in my model of the Taj Mahal. The main mausoleum structure of this modern wonder is a cube with each corner cut off at a 45-degree angle to form an unequal octagon. Figure 6-20 shows one corner section of the model. Based on the scale I was using, the $(7,7,10)$ near triple was perfect to create the angled section at each corner.

Figure 6-20: An angled wall in the Taj Mahal model

MORE OPTIONS WITH JUMPER PLATES

Pythagorean triples and near triples also work when you multiply all the numbers in a triple by a whole number, like 2, or when you halve them. For instance, take the triple (3,4,5) and multiply all the numbers by 2, and you get another triple, (6,8,10), as we saw in the Spring Lantern Festival set. Similarly, if you take the near triple (7,7,10) and halve all the numbers, you get (3.5,3.5,5). There's no reason to think this halved near triple wouldn't work to create an angled wall with LEGO, but how do you create a triangle that has sides that are 3.5 studs long? Using jumper plates, of course! Figure 6-21 shows a 45-degree angle created with the (3.5,3.5,5) near triple and jumper plates to get the extra half stud of length on each side.

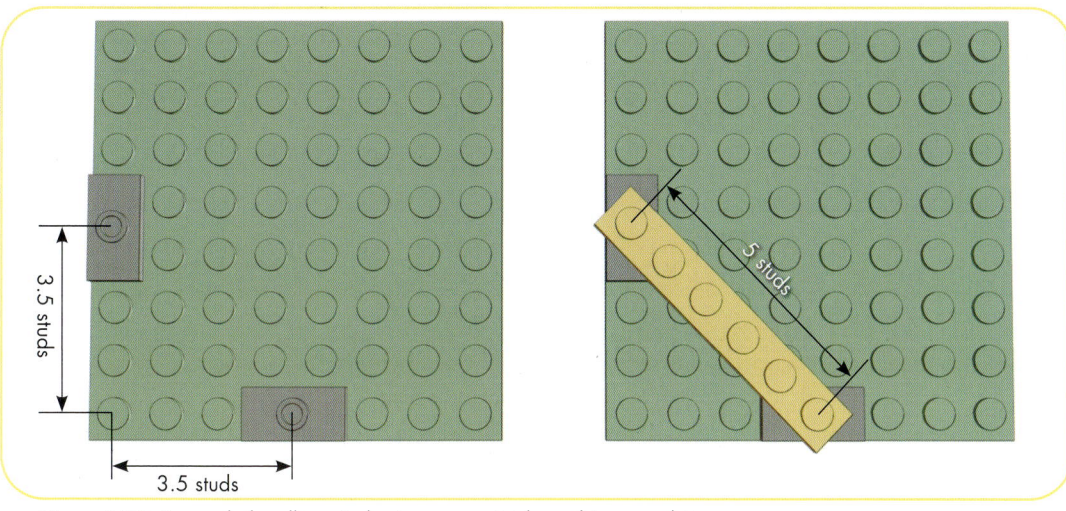

Figure 6-21: An angled wall created using a near triple and jumper plates

Figure 6-22: The crown of the Chicago Tribune Tower in LEGO form

Combining near triples with half-stud increments gives you more options for building 45-degree walls. One combination I've found particularly useful is (8.5,8.5,12), which is half of the near triple (17,17,24). I employed this and other near triples to achieve the various angled walls of my model of the Tribune Tower in Chicago (Figure 6-22).

The chamfered corners of the main tower of this neo-Gothic skyscraper used the near triple (5,5,7). The highly ornate crown of the building features two octagonal levels along with flying buttresses that are reminiscent of Rouen Cathedral in France (which inspired the building's design). I used two near triples for the octagonal sections: (7,7,10) and (8.5,8.5,12), the latter requiring jumper plates.

OTHER ANGLES

Near triples aren't just for creating 45-degree angles. For instance, (4,7,8) and (4,8,9) are helpful near triples that produce other angles. I used (4,7,8) to connect the flying buttresses on my Tribune Tower model from the octagonal crown to the main walls of the tower below. Figure 6-23 shows a top view of the model, highlighting its various near triples.

Each of the eight flying buttresses projects out from one of the corners of the octagonal crown and meets the main tower at approximately a 30-degree angle. With the scale I was using, the (4,7,8) near triple was perfect for achieving this angle.

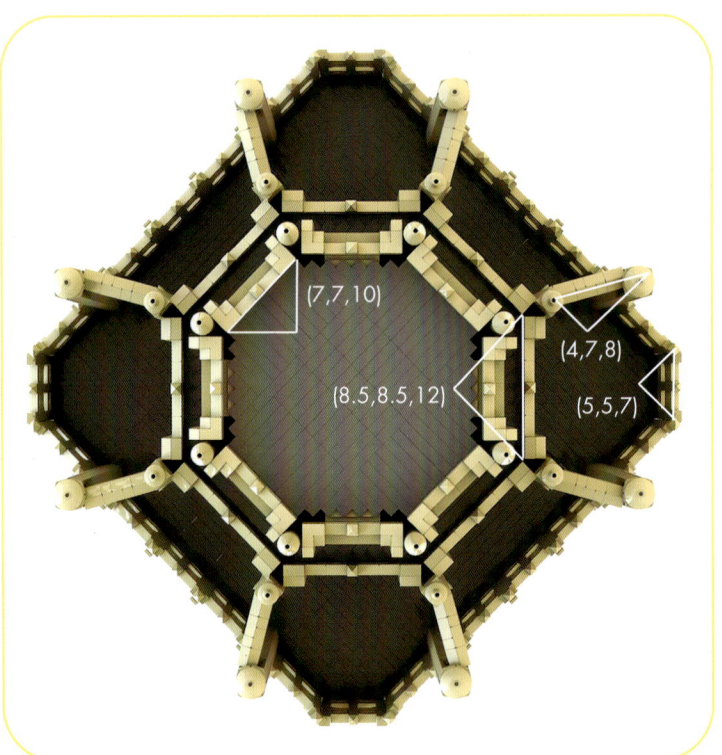

Figure 6-23: The different near triples used to create angled walls in the model of the Chicago Tribune Tower

MORE ANGLED BUILDING TECHNIQUES

The techniques we've seen so far create angled walls by placing elements along the hypotenuse (angled side) of a right triangle. For this to work, the length of the hypotenuse has to be a whole number of studs, and this limits our options to Pythagorean triples and near triples. However, there are a few other angled building techniques where we don't place any elements along the hypotenuse and can therefore disregard its length. This opens up quite a few other possibilities.

THE MIRRORED HYPOTENUSE

The *mirrored hypotenuse technique* takes advantage of the angle that occurs when you place two mirror images of the same right triangle next to each other, such that they share a hypotenuse. Let's start with an arbitrary right triangle—say, the sides that form its right angle are 6 studs and 2 studs long. The length of the hypotenuse would be $\sqrt{(6^2 + 2^2)} = 6.32$ studs, which isn't a whole number. We can't place a LEGO element directly along this side, but if we mirror the right triangle along the hypotenuse, we can create an angled wall by placing LEGO elements along the other two sides of the second triangle. We'll need to use hinge plates to join the two mirrored right triangles together, as shown on the left of Figure 6-24.

WEDGE PLATES

A more common application of the mirrored hypotenuse technique uses *wedge plates* to create the two mirrored triangles, as seen on the right of Figure 6-24. These are plates that already have angled sides. The example on the right of the figure uses two 6×3 wedge plates (#54383 and #54384) to create the same mirrored triangles shown on the left.

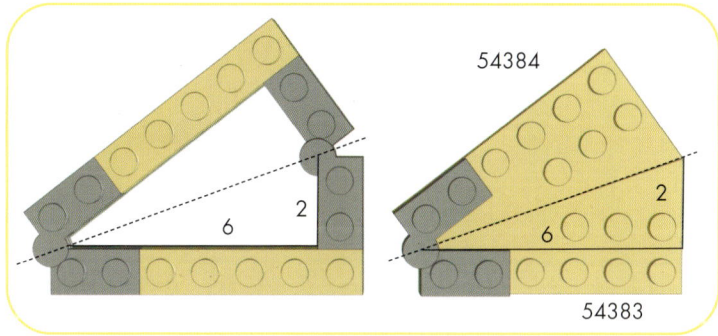

Figure 6-24: The mirrored hypotenuse technique without (left) and with (right) wedge plates

This technique is used extensively in LEGO's Boutique Hotel set, where the floor and roof sections feature hypotenuse-to-hypotenuse 6×3 wedge plates to create the angled portion of the building. For example, Figure 6-25 shows the floor of the hotel's second story.

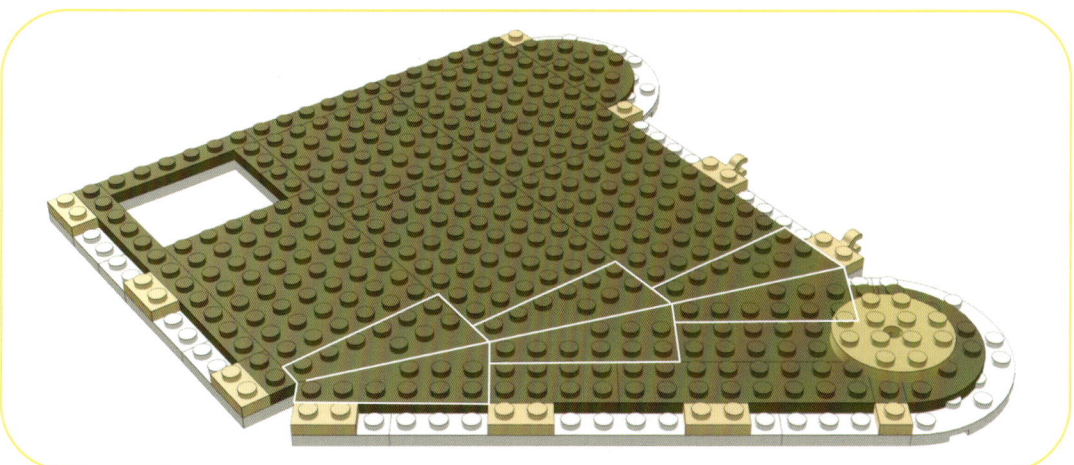

Figure 6-25: The mirrored hypotenuse technique used in the Boutique Hotel set

The floor and roof sections of the Boutique Hotel are built using regular plates and wedge plates, and yet somehow their angled sides line up perfectly with the angled walls of the building that are created, as we discussed earlier in the chapter, using (3,4,5) triangles. Figure 6-26 shows how they fit together. Is this a coincidence? Not really.

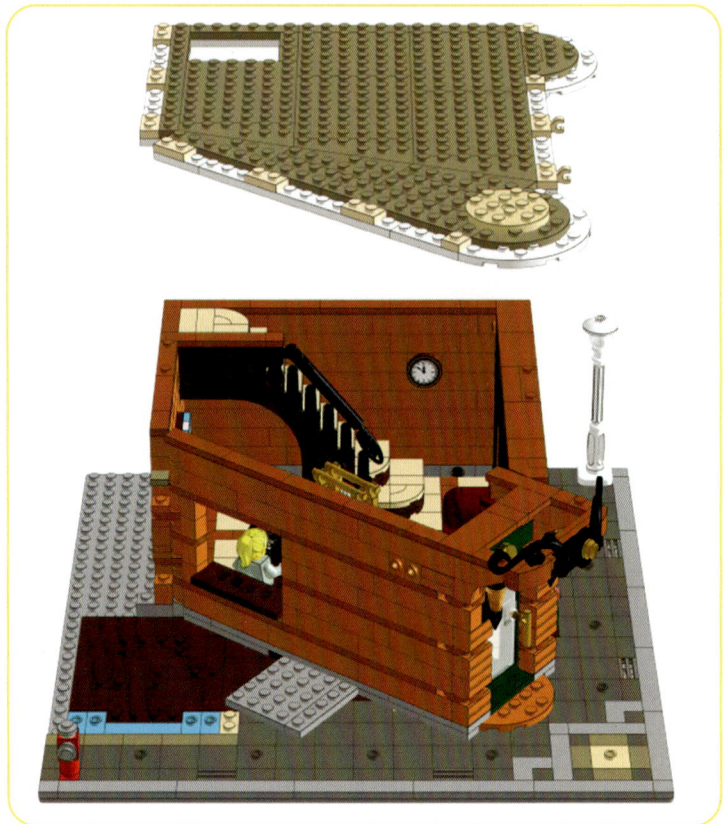

Figure 6-26: The different floors of the Boutique Hotel

Each wedge plate has a right triangle with a tangent of 2/6 = 0.333, and the arctan of that is 18.4 degrees. It makes sense that two of these wedge plates would give us a combined angle of 36.8 degrees (18.4 × 2), which matches the angle created using the (3,4,5) triple. Figure 6-27 shows how they compare.

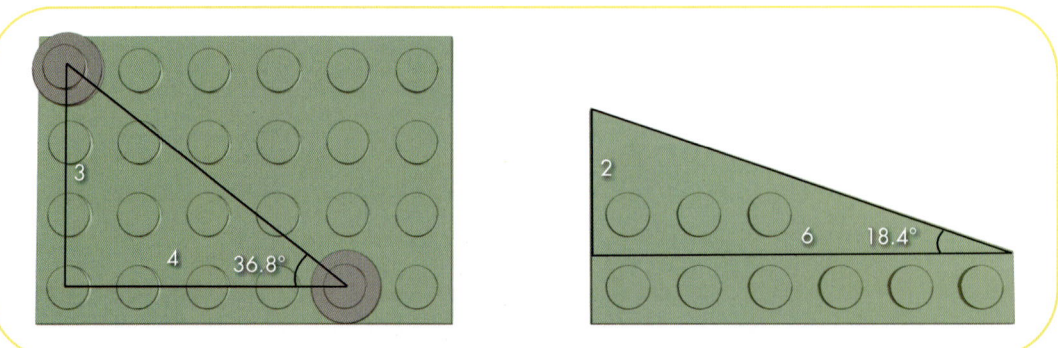

Figure 6-27: Angles created by a (3,4,5) triangle and 6×3 wedge plates

Table 6-2 shows all the angles that can be created using pairs of the various LEGO wedge plates.

TABLE 6-2: Common Wedge Plates

Dimensions	Part numbers	Angle (degrees)
2×2	24299, 24307	53.13
3×2	43722, 43723	36.87
4×2	41769, 41770	28.07
6×2	78443, 78444	18.92
6×3	54383, 54384	36.87
6×4	48205, 48208	53.13
8×3	3544, 3545	28.07
12×3	47397, 47398	18.92

PYTHAGOREAN QUADRUPLES

It's also possible to extend the mirrored hypotenuse technique to two unequal right triangles, where the hypotenuse of one triangle becomes one of the non-hypotenuse sides of the second triangle. The resulting quadrilateral (four-sided shape) would have to satisfy the equation for a *Pythagorean quadruple*, which is $a^2 + b^2 + c^2 = d^2$. The simplest Pythagorean quadruple is (1,2,2,3). Multiply all numbers by 2 and you get (2,4,4,6), which is a Pythagorean quadruple as well. Figure 6-28 shows how to build a (2,4,4,6) quadruple using LEGO plates and hinge elements.

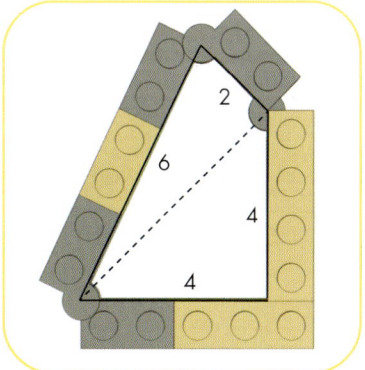

Figure 6-28: An application of a Pythagorean quadruple

The neat thing about Pythagorean quadruples is that they can be used to place elements at an angle not just in two dimensions but in three dimensions as well. Figure 6-29 extends the (2,4,4,6) quadruple into 3D space.

Realizing this 3D quadruple requires a different type of hinge element, the 1×2 hinge brick assembly consisting of #3937 and #3938. With two of these hinge bricks placed on 2×2 turntables at different heights, we can attach a 2×8 plate at an angle in three dimensions.

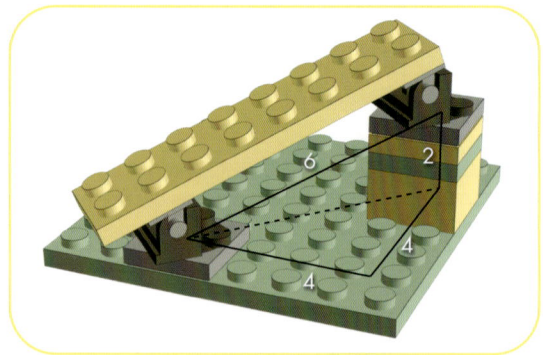

Figure 6-29: The Pythagorean quadruple in 3D space

SWITCHED DIAGONALS

The diagonal distance between two studs on a plate may not be a whole number of studs. But if you take a rectangular plate, the distances between the two pairs of studs at opposite corners (from 1 to 3 and from 2 to 4 on the left of Figure 6-30) are exactly the same. You can rotate the plate and attach it such that corners 2 and 4 line up with where corners 1 and 3 normally would be, a trick known as the *switched diagonals technique*. This is shown on the right of Figure 6-30. Once again, you need to use 1×1 plates as spacers.

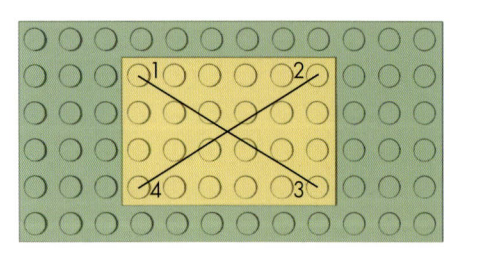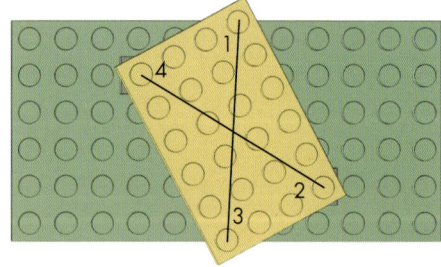

Figure 6-30: The switched diagonals technique illustrated

If you think of 1–3 and 2–4 diagonals as the hypotenuses of two identical right triangles that are mirrored, all you're doing in the switched diagonal technique is rotating one of the triangles so that its hypotenuse lines up with that of the other triangle. The angle that the mirrored right triangle contains at corner 4 is arctan(3/5) = 30.96 degrees, but the total rotation required to line up the two hypotenuses is twice that angle. This is because you first have to rotate line 2–4 by 30.96 degrees to get it to be horizontal and then by another 30.96 degrees to get it to line up with line 1–3. That makes the total angle of rotation 61.92 degrees.

Table 6-3 shows the angles you can achieve by applying the switched diagonals technique to different sizes of plates. The closest you can get to a 45-degree rotation is by using a 4×8 plate.

TABLE 6-3: Switched Diagonal Possibilities

Plate	Part number	Angle (degrees)
2×3	3021	53.13
2×4	3020	36.87
2×6	3795	22.62
2×8	3034	16.26
2×10	3832	12.68
2×12	2445	10.38
4×6	3032	61.93
4×8	3035	46.39
4×10	3030	36.87
4×12	3029	30.51
6×8	3036	71.08
6×10	3033	58.11
6×12	3028	48.89

This switched diagonals technique can be extended to include hinge elements, as shown in Figure 6-31. In this case, the technique looks a little different because the diagonals that are switched go all the way to the corners of the plates rather than the centers of the corner studs. This has to be accounted for when we calculate the angle of rotation that can be achieved.

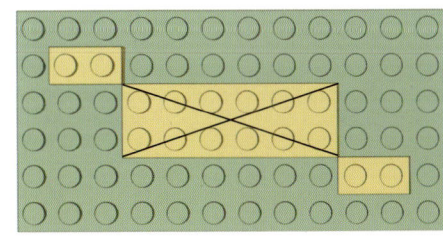 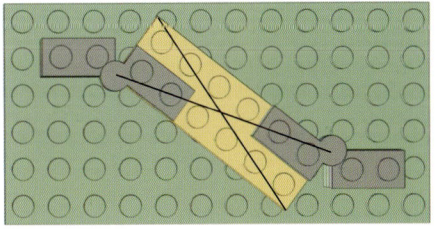

Figure 6-31: The switched diagonals technique using hinge plates

SUMMARY

We've seen how the square grid of the LEGO system doesn't have to limit you to building structures with 90-degree corners. It's possible to create other angles if you're mindful of the underlying math governing the right triangles that are created by the angled walls. We'll extend the concept of angled walls further to create closed polygons and then round shapes in Chapter 7.

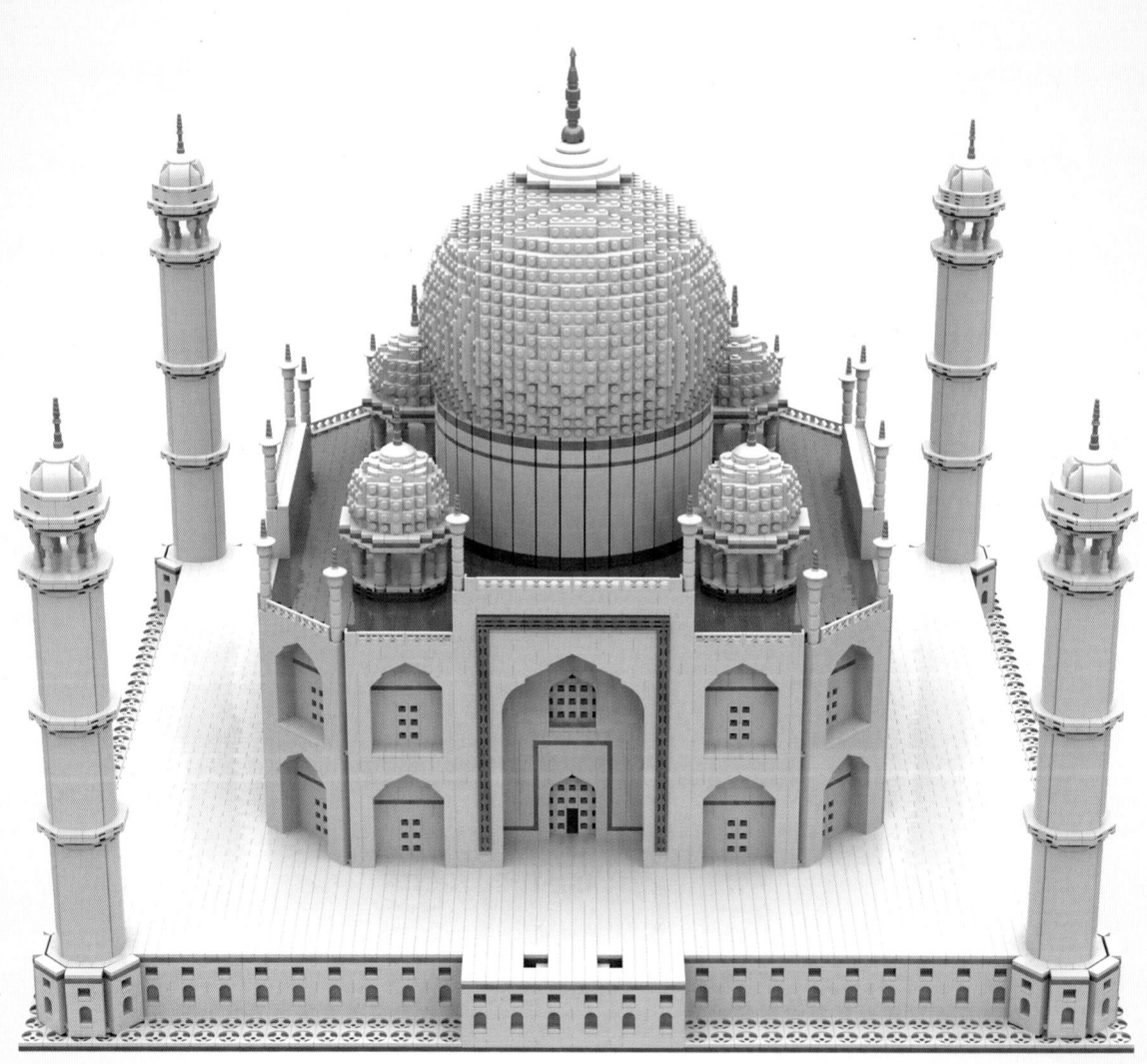

ROUND SHAPES

LEGO isn't a medium inherently suited to building round shapes. After all, it's a system based on a regular square grid that consists primarily of rectangular elements. Despite this, there are many wonderful LEGO creations that feature organically curving forms. In this chapter, we'll look at some techniques for building round shapes like cylinders, circular walls, domes, and even spheres.

We'll also consider what's possible using LEGO's limited range of rounded elements, but our main focus will be on how to use regular bricks and plates to create the best possible approximation of a curve. The result is never perfect, and the limitations of the LEGO medium are always apparent in the jaggedness and gaps that you may see, especially when you look at your model up close. The trick is to create a convincing *illusion* of a round shape, at least when viewing your model from a few steps back.

CURVED LEGO ELEMENTS

The LEGO catalog includes a selection of round bricks, plates, and tiles. Specialized pieces are also available for creating round walls, hollow cylinders, and domes (see Figure 7-1). These pieces come in only a few different sizes, however, which limits the dimensions of the round structures you can build with them. For example, the biggest dome you can create using dome pieces has a diameter of 10 studs.

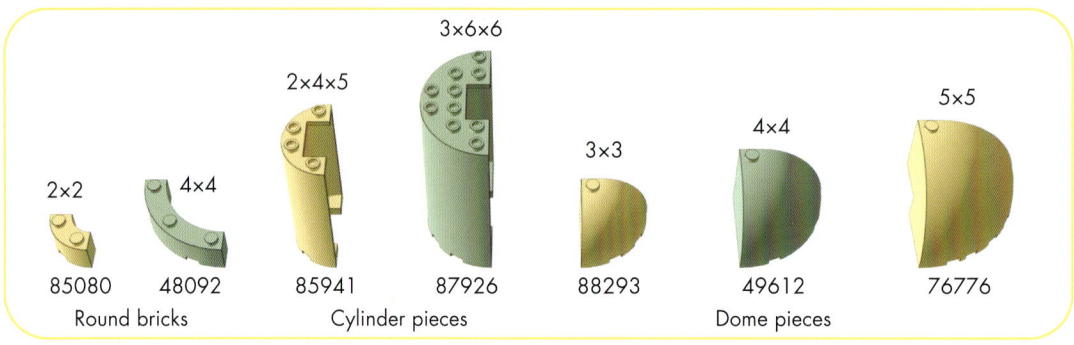

Figure 7-1: A selection of LEGO round bricks, cylinder pieces, and dome pieces

Another category of rounded LEGO elements is the curved slopes we first saw in Chapter 3. We can think of the sloped portion in a curved slope as the arc of a circle, and so technically, if we put enough of these pieces together, we should be able to form a circle. The tricky part is that we have to precisely position these pieces in different orientations. For that, we first need to use SNOT techniques to create a core that has studs facing in four different directions. We can then attach the curved slope pieces to the outside of this core to create a cylinder.

This approach is much more flexible than using LEGO's dedicated cylinder pieces, since you can achieve a wider variety of diameters (see Figure 7-2).

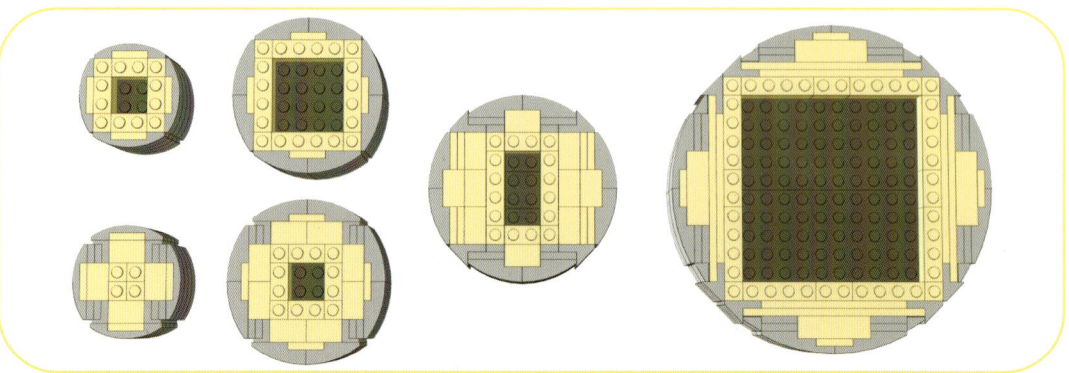

Figure 7-2: Cylinders created by attaching curved slope pieces to a SNOT core

The usual 6:5 ratio of sideways building applies, so the cylinder can be built up in height increments of 2 studs (or 5 plates).

The smallest of these cylinders consists of 1×4 curved slope pieces (#93273) attached on all four sides of a 4×4 square SNOT core. A 2×4 plate runs down the middle of each face of the core, filling in the gap on the underside of each slope piece and locking all the slopes into place (see Figure 7-3).

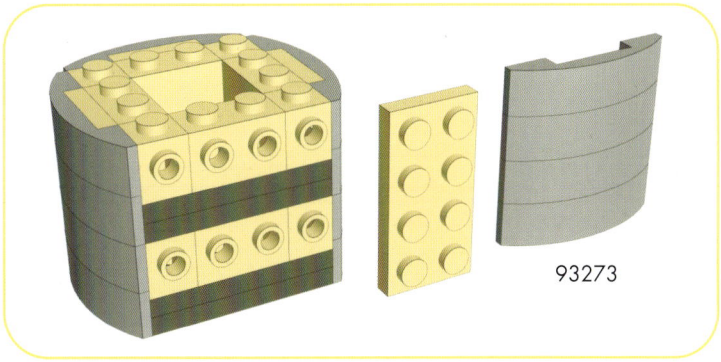

Figure 7-3: The smallest of the cylinders used for the minarets of the Taj Mahal

The result has a diameter of just under 6 studs, the perfect size for the minarets on my Taj Mahal MOC (Figure 7-4). These SNOT cylinders rely on curved slopes for their rounded form. For the rest of the chapter, however, we'll consider what round shapes are possible using primarily square and rectangular elements.

Figure 7-4: Taj Mahal minarets created by stacking cylinders

BUILDING ROUND WALLS

Round walls are an eye-catching addition to architectural builds. For example, you can use them to create a turret of a castle or a *tholobate* (also known as a *drum*), the cylindrical portion of a building beneath a dome. Let's look at a few different techniques for building round walls.

BENDING BRICKS

A brute-force approach to building a rounded wall is to build a straight wall using 1×2 bricks in a staggered bond pattern and then bend it to form a circle (see Figure 7-5). The longer your wall, the more flex it will have, making it easier for you to bend it into a complete circle. The number of 1×2 bricks needed in each layer to build a stable round wall tends to be around 72. The technique works with plates as well.

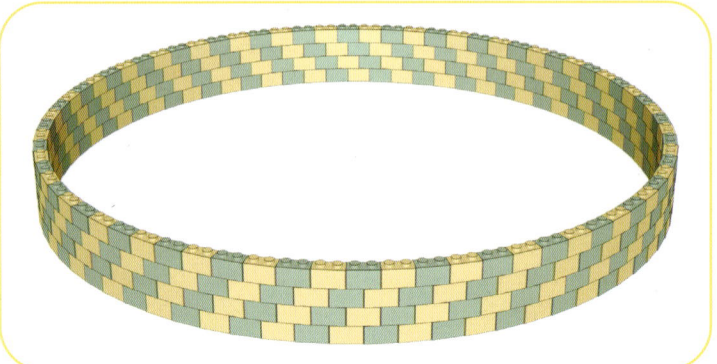

Figure 7-5: Bending a LEGO wall to create a round shape

LEGO artist Jeff Sanders specializes in art created using various *brick-bending* techniques like this. His amazing creations are made not just by bending LEGO walls into simple circles but also by interconnecting multiple curved wall segments to form intricate patterns. Strictly speaking, most brick-bending techniques are illegal because you're using LEGO elements in ways they aren't intended to be used and sometimes subjecting them to undue stress and possible damage. Another downside to this technique is that it can't easily be replicated digitally in Studio.

MIXING RECTANGULAR AND ROUND BRICKS

Another approach to building round walls is to alternate between rectangular bricks (1×3 or 1×2 bricks work well) and round 1×1 bricks in each layer. This technique is similar to brick bending, but the round bricks act like hinges, allowing the wall to be bent more easily (and legally!) into a circle. As illustrated in Figure 7-6, the method allows round walls to be built with smaller diameters than are possible with brick bending.

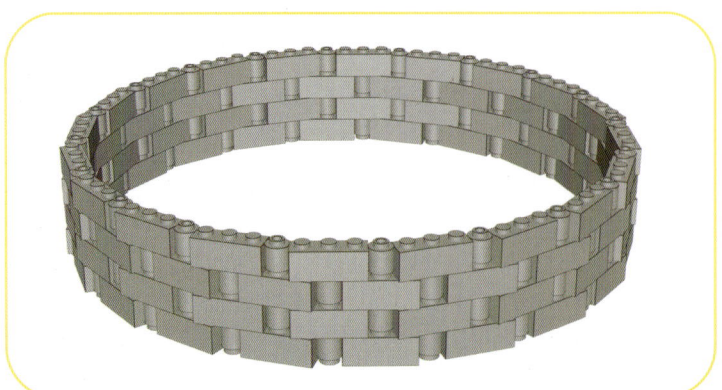

Figure 7-6: A round wall created by mixing 1×1 round bricks with regular 1×3 bricks

One downside to this technique is that the texture of the wall is uneven due to the 1×1 round bricks. However, it's possible to cover the outer surface of the wall with tiles to hide the round bricks, as seen in Figure 7-7.

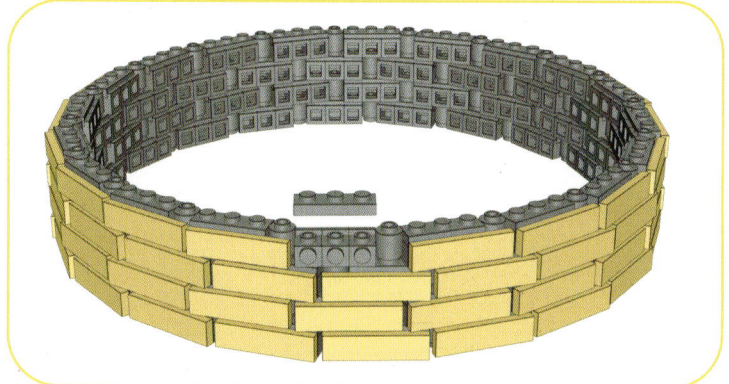

Figure 7-7: Using headlight bricks instead of 1×3 bricks allows 1×4 tiles to be attached to the outer surface of the wall.

Of course, adding the tiles requires SNOT building. Here I've replaced each 1×3 brick with three headlight bricks with their top studs facing out. These headlight bricks are joined together by a 1×3 plate to reach the full height of 1 brick. Meanwhile, the 1×4 tiles on the face of the wall mimic the look of real bricks, which should work great for castle builds.

USING HINGE PLATES

In Chapter 6, we discussed how to use LEGO hinge plates to create angled walls. If we extend this concept further, we should be able to create a continuous chain of angled wall segments, all of the same length, that come together to form a closed shape—that is, a *regular polygon*. Give that polygon enough sides (say, 20 or more), and it should pass for a circular wall.

SIMPLER POLYGONS

Let's first take a step back and see what it would take to build a simpler polygon, like a hexagon or an octagon. Then we'll apply what we've learned to making a circle. The challenge here is that it isn't enough to just build a polygonal structure using hinge pieces; we also need a way to incorporate it into a LEGO model by firmly attaching it to a baseplate. To make a good connection, we need at least one pair of opposite sides of the polygon to line up with the LEGO grid, even if the rest of the sides don't. For this reason, it's much easier to create polygons with an even number of sides.

Whether a pair of sides will align depends on the mathematical relationship between the length of each side of a regular polygon and the length of its *apothem* (the distance from the center of a polygon to the center of one of its sides—see the "Polygon Geometry" box for more information). What we need is a polygon where the side length is a whole number of studs, and where twice the apothem (the distance between two opposite sides) is also a whole number of studs—or close enough to a whole number, anyway.

POLYGON GEOMETRY

A *regular polygon* is a closed shape that has *n* sides of equal length, where *n* is 3 or greater. The point where any two adjacent sides intersect is called a *vertex*, and the line from a vertex to the center of a polygon is called the *radius*. As illustrated here, the angle between two adjacent radii is called the *central angle*. An *n*-sided polygon's central angle is equal to 360 degrees divided by *n*.

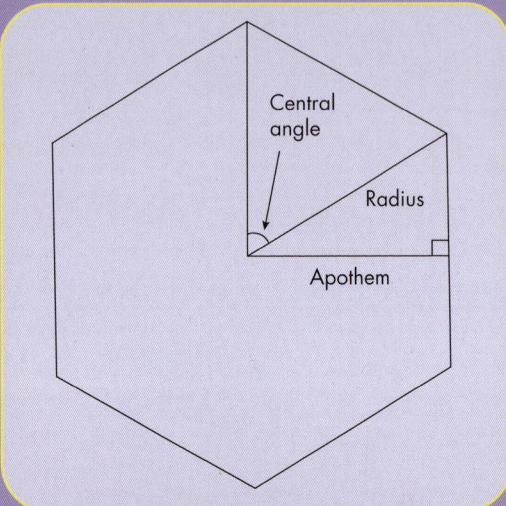

The *apothem* is a line drawn from the center of a polygon to the center point of one of its sides. This line is perpendicular to the side, and twice the apothem gives you the perpendicular distance between two opposite sides. A polygon's apothem is shorter than its radius, but these two lines combine with half a polygon's side to form a right triangle. So calculating the radius or apothem based on a polygon's side length is a simple matter of trigonometry. Or you can use an online tool to make the calculation for you!

Luckily, there are plenty of online apothem calculators that can help you figure out what side lengths will work for the polygon you're trying to build. Say you want to build a hexagon. Start plugging in different side lengths (2, 3, 4, 5, . . .) and you'll see that the shortest side length that yields (close to) a whole number for twice the apothem is 7 studs. The apothem for this hexagon is approximately 6 studs, making the distance between any pair of opposite sides about 12 studs. For an octagon, the smallest side length that works is 5 studs, and the distance between the opposite sides is, once again, around 12 studs. Figure 7-8 shows these two polygons, built with a mix of hinge plates and regular plates. Notice how the top and bottom sides of each polygon align with the LEGO grid.

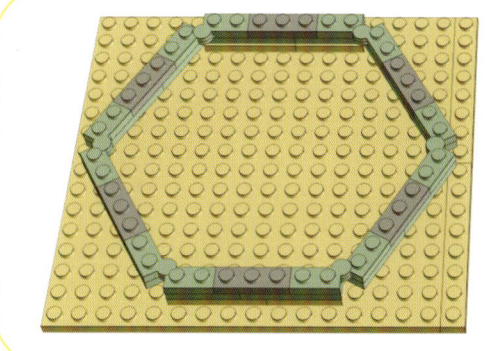

Figure 7-8: The smallest hexagon (left) and octagon (right) that can be securely attached to a LEGO baseplate

FROM POLYGONS TO CIRCLES

The same principle applies if we're trying to approximate a circle: one pair of opposite sides needs to align with the LEGO grid. To appear round, the polygon needs to have many sides, and each side should be quite short. The shortest we can go is a side length of 2 studs, achieved with two layers of 1×4 hinge plates. The hinge locations are offset by 2 studs between the layers, as shown in Figure 7-9.

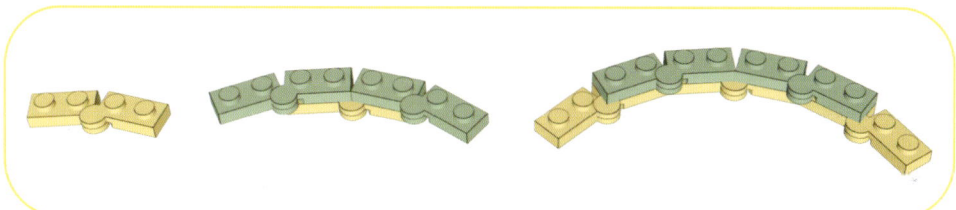

Figure 7-9: Building a round wall using two layers of hinge plates

We're holding the side length constant, but the number of sides (that is, the number of hinge plates) can vary based on the angle of each hinge. How many sides will work? A nice thing about a regular polygon with a large number of sides is that we can start thinking of its apothem as the radius of a circle, and twice the apothem as the circle's diameter. This lets us bypass apothem calculations and instead use the simpler formula $C = \pi d$, where C is the circle's circumference and d is its diameter. The circumference here (in studs) is the number of sides times 2, and dividing that by π gives us the diameter. If the diameter is close enough to a whole number, we'll be able to attach the round wall to a baseplate. Some side counts that work are 22 (diameter of 14 studs; see the left of Figure 7-10), 28 (diameter of 18 studs; see the right of Figure 7-10), 30 (diameter of 19 studs), and 36 (diameter of 23 studs; see the left of Figure 7-11).

Figure 7-10: Round walls with diameters of 14 (left) and 18 (right) studs

Figure 7-11 shows how I used a round wall with a diameter of 23 studs to create the tholobate in my model of the Taj Mahal. I included a second ring of hinge plates near the top to better hold the individual sides together.

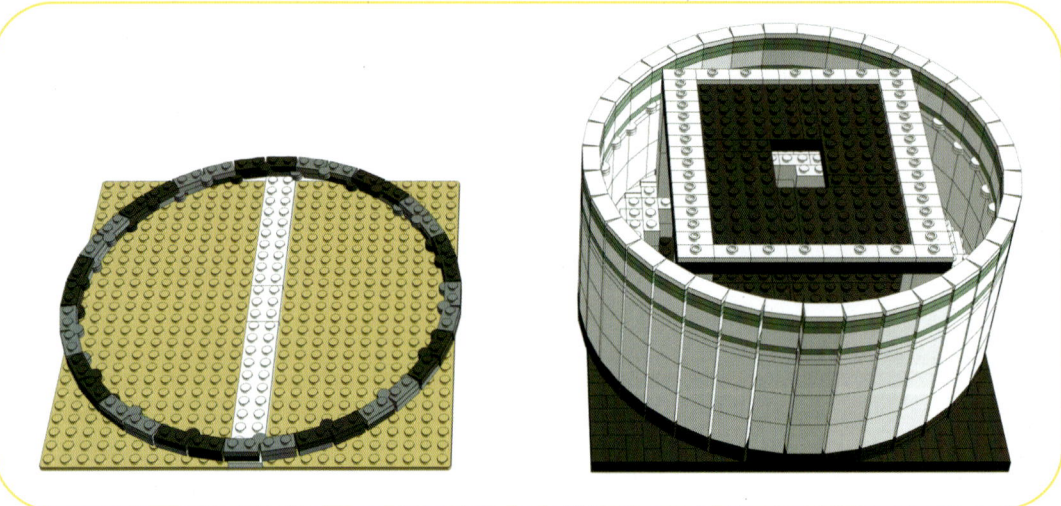

Figure 7-11: A round wall used for the tholobate (or drum) of the Taj Mahal

INTERNAL SUPPORT STRUCTURES

One issue you may encounter when using this technique is that it isn't easy to get the wall to keep its round shape. When there are just two connection points to the base, the rest of the hinged wall segments are free to move. There are ways to create an internal support framework using Technic elements, but it's simpler to build an inner wall using regular bricks as close as possible to the round wall. You can then use SNOT techniques to attach cheese slopes, curved slopes, and other elements to the inner wall, filling in the gaps and helping the outer wall retain its shape.

DIGITAL BUILDING TIP

In Chapter 6, we discussed how to rotate elements like hinge plates for angled building in Studio. The same technique applies to building a round wall digitally. The key is to make sure each hinge plate is rotated by just the right amount or else the two ends of the wall won't line up correctly and complete the circle. The angle of rotation necessary is the same as the central angle of the polygon. As mentioned in the "Polygon Geometry" box, this angle is equal to 360 divided by the polygon's number of sides. For a round wall with 28 sides (diameter of 18 studs), for example, the angle would be 360/28 = 12.85 degrees. It's best to enter this angle manually (as described in Chapter 6). Each hinge assembly in the round wall needs exactly the same angle of rotation, so once you've rotated one hinge plate, copy and paste the hinge assembly (both halves together) as many times as needed to complete the circle.

BUILDING A SPHERE

A classic challenge in the LEGO world is using regular bricks and plates to get as close as possible to building a sphere. Spheres (or partial spheres) can have a wide range of applications in LEGO builds to create everything from a globe or a soccer ball to the dome of a building like the Taj Mahal.

STACKED BRICKS

One way to build a LEGO sphere is to stack layers of regular bricks, all with the studs facing up. Each layer is essentially an approximation of a circle. Start with the middle layer, the widest part of the sphere, and build out symmetrically from there, creating successively narrower "circles" above and below until you reach the top and bottom of the sphere. The question then becomes: For each layer of the sphere, how do you best approximate a circle?

APPROXIMATING A CIRCLE

It turns out LEGO builders aren't the only ones facing a dilemma in trying to create round forms out of square building blocks. The *Minecraft* community, too, has to navigate this challenge, and they've developed some resources that are readily adaptable to LEGO. For example, *Minecraft* enthusiasts often use a *circle chart* like the one shown in Figure 7-12.

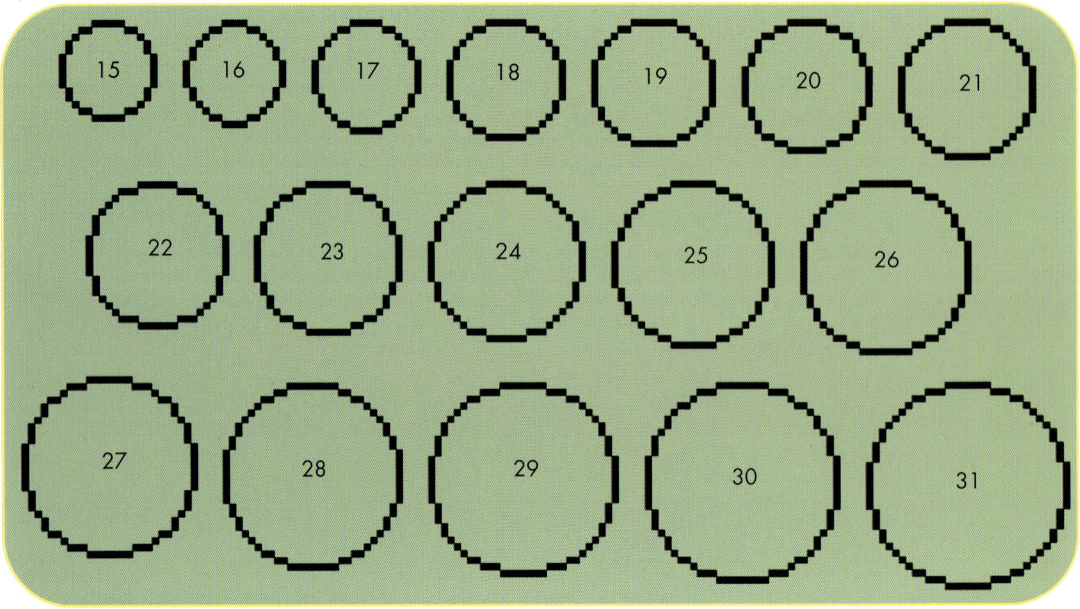

Figure 7-12: A *Minecraft* circle chart

A circle chart shows the best way to approximate a two-dimensional (2D) circle of various diameters by placing blocks (or pixels) in a square grid. The bigger the diameter, the more convincing is the illusion of the round shape. Consulting a chart like this, you could determine the ideal footprint of each layer of your LEGO sphere. Or, if that sounds a bit tedious, you can use Plotz (*https://www.plotz.co.uk*), an online 3D modeling assistant for *Minecraft*, to automate this process. With its Sphere tool, you just enter the sphere diameter you need, and presto, a sphere is generated for you, with 3D and 2D views showing how to build the sphere layer by layer (see Figure 7-13).

Getting from a Plotz model to an actual LEGO sphere still requires some thought. The sphere generator shows individual *Minecraft* blocks, the equivalent of 1×1 LEGO bricks, but you can't simply stack a bunch of 1×1s the way you see in the 3D view. You need to convert the 1×1s into longer bricks to form an interlocked structure that holds together well. You also need to make the walls of the sphere at least 2 studs deep, allowing each successive layer to rest on the layer immediately below it. Alternating the orientation of the bricks from one layer to the next, as discussed in Chapter 3, will yield the strongest result.

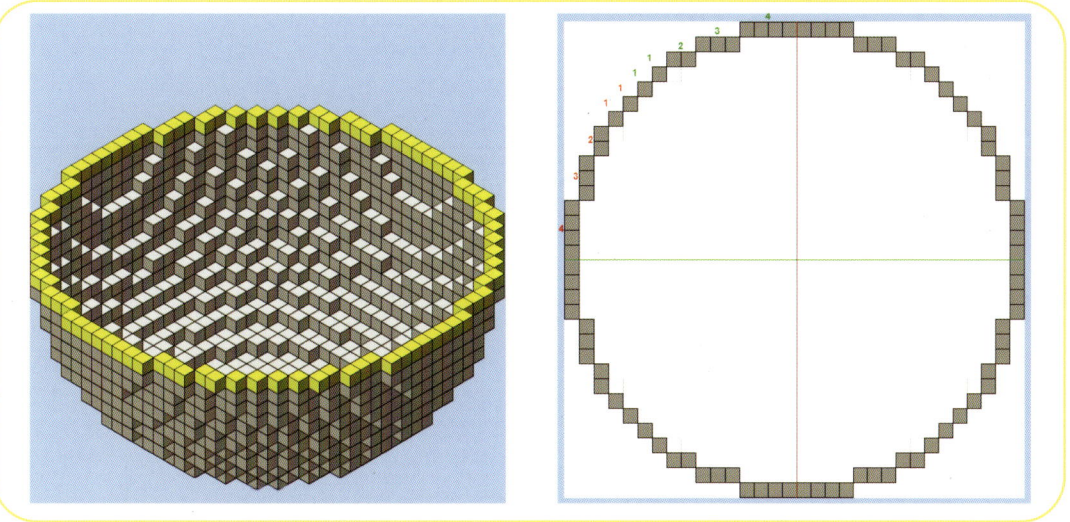

Figure 7-13: A sphere generated by Plotz, with a 3D and 2D view of the highlighted layer

MAINTAINING THE CORRECT PROPORTIONS

There's one last hitch. *Minecraft* blocks are perfect cubes, but 1×1 LEGO bricks, of course, are not. If you replace all the blocks in the Plotz model with LEGO bricks, your sphere will end up a little taller than it is wide—just like a 1×1 brick. Thankfully, Plotz also has an Ellipsoid tool, and you can use that to compensate for LEGO proportions. This tool lets you set the height, width, and depth of the 3D shape separately, as opposed to the single Size controller in the Sphere tool.

The key is to set the width and depth of the ellipsoid to your desired LEGO sphere diameter (in studs), and then set the height to five-sixths that diameter. For example, the left half of Figure 7-14 shows a Plotz ellipsoid with a width and depth of 24 units and a height of 20 units. The right half of the figure shows this ellipsoid translated into LEGO.

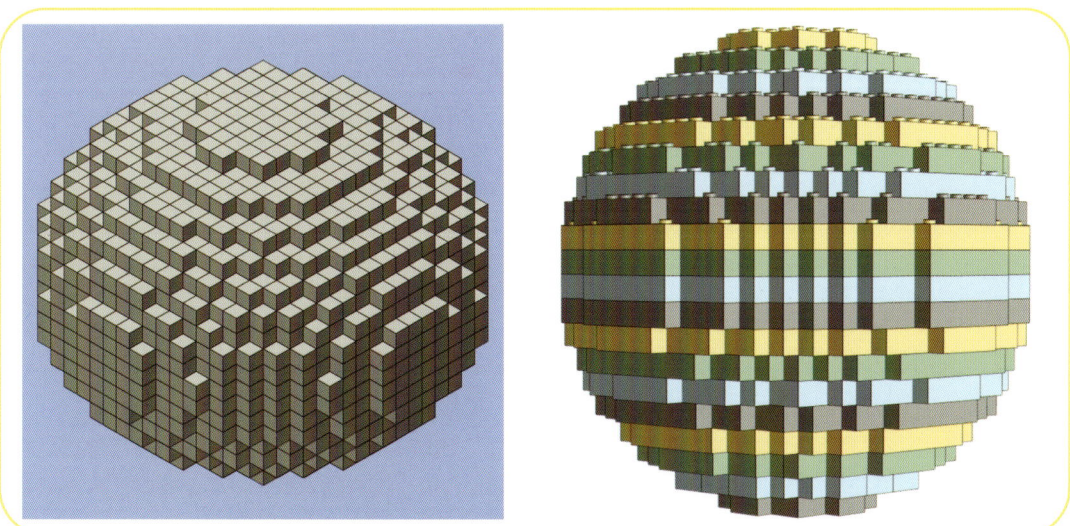

Figure 7-14: A Plotz ellipsoid (left) and the equivalent sphere built by stacking LEGO bricks (right)

Since a 1×1 brick has a height-to-width ratio of 6:5, the LEGO equivalent of a 5:6 ellipsoid works out to be perfectly spherical. The ratios cancel each other out.

STACKED PLATES

A sphere made from stacking bricks inevitably looks a little blocky. If we stack thinner LEGO plates rather than bricks, is it possible to smooth out the curves? To find out, we can go back to the Plotz ellipsoid generator and adjust the dimensions. Since a plate is one-third the height of a brick, a 1×1 plate has a height-to-width ratio of 2:5. The ellipsoid therefore needs to have the inverse ratio of 5:2. For example, if it's 24 units wide and deep, it should have a height of 60 units. Figure 7-15 shows the resulting LEGO sphere.

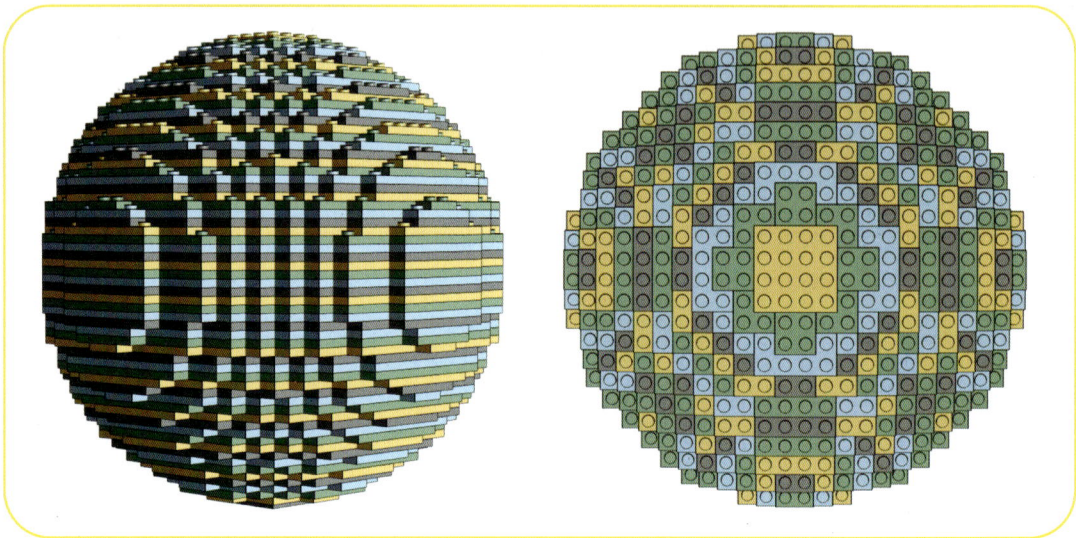

Figure 7-15: A sphere built by stacking LEGO plates

Looking at the sphere from the side, there's a definite improvement. The curves are less jagged thanks to the smaller height gradations achieved by using plates. But viewed from the top, the curves, once again, appear blocky. Plates still have the same basic footprint as bricks, so from the standpoint of width and depth, stacking plates is no better than before.

Furthermore, another downside to building spheres via simple stacking, whether it be bricks or plates, is that the undersides of the pieces will be visible when you view the sphere from below. Ideally, a sphere should have a uniform appearance no matter how it's oriented. To achieve that uniformity and smooth out the curves in all three dimensions, we need to turn to SNOT building.

LOWELL SPHERES

In 2002, Bruce Lowell revolutionized the construction of LEGO spheres by breaking them down into six identical curved panels built using LEGO plates. The panels are then joined together around an interior SNOT cube with studs in all six directions. Each panel is longer than it is wide, allowing the panels to interlock perfectly without any visible gaps and form a wonderfully smooth and symmetrical sphere. The result, known as a *Lowell sphere*, after its creator, is shown in Figure 7-16.

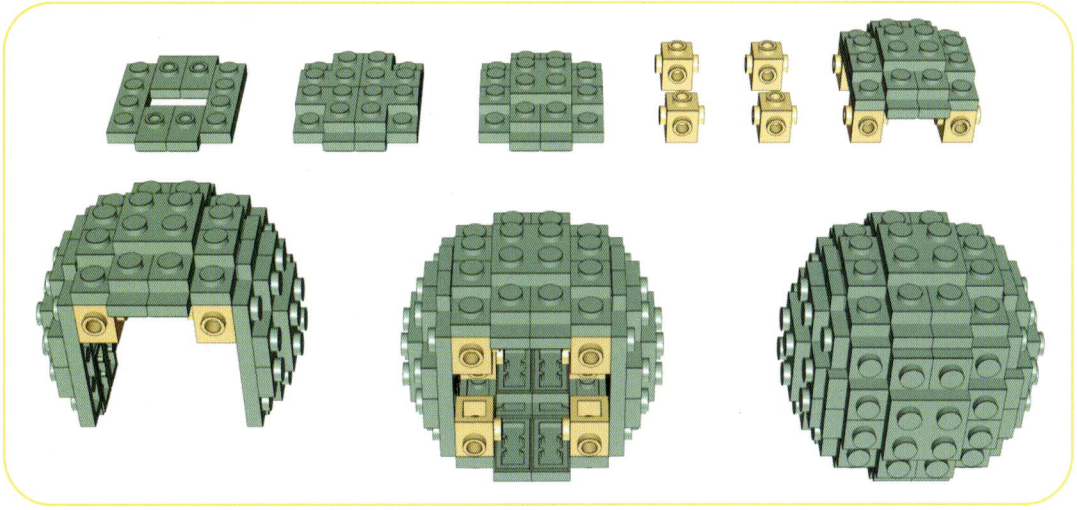

Figure 7-16: A basic Lowell sphere

Lowell's original sphere had a diameter of 6.8 studs and an inner 4×4×4 core. Over time, the technique has been generalized to other dimensions. It's also found applications in sculpting complex shapes other than spheres, as we'll discuss in Chapter 9.

BRAM'S SPHERE GENERATOR

Much of the credit for the Lowell sphere's expansion goes to Bram Lambrecht and his online tool Bram's Sphere Generator (*https://lego.bldesign.org*). This tool can help you create a Lowell sphere with any diameter you need. Just enter the diameter (in increments of 0.2 studs) and tweak a few settings (such as whether to use jumper plates for even finer detail), and you're ready to save an LDraw file of a sphere, grouped into six panels, that can be imported into Studio. Figure 7-17 shows an example, with an individual panel on the left and the full sphere on the right.

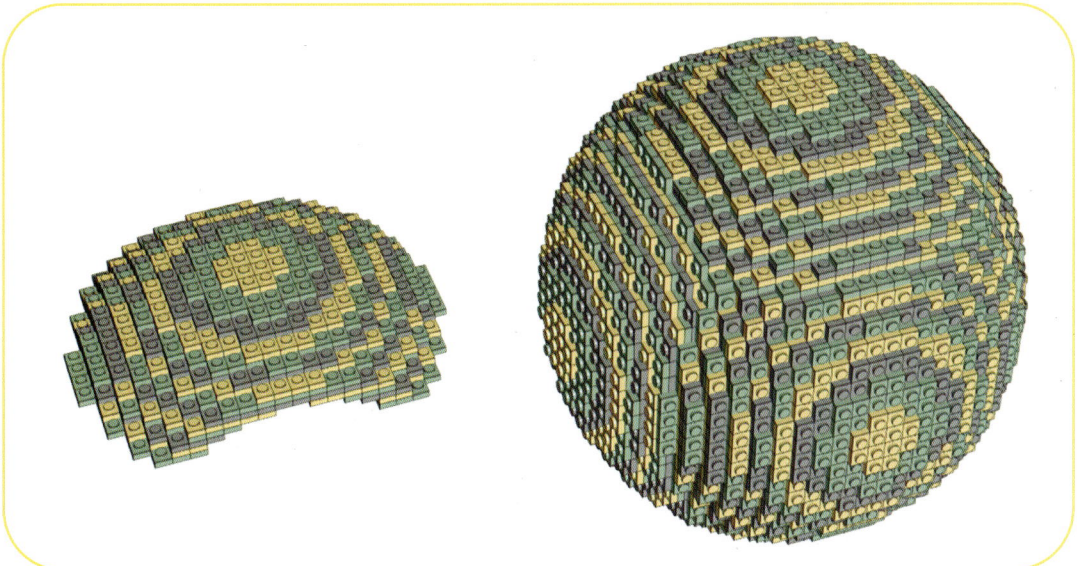

Figure 7-17: A bigger Lowell sphere

The resulting sphere is made entirely from 1×1 plates (plus 1×2 jumper plates if you've enabled that option). Therefore, as with consulting Plotz, there's still some manual work involved in realizing the sphere. You have to replace the 1×1 plates with bigger pieces that can interlock to hold each of the six panels together as a unit, ensuring that the seams between plates don't line up between successive layers of the panel. The Sphere Generator's Use Alternating Layer Colors option helps with this process, making it easy to see which layer is which.

The LDraw file also doesn't include a SNOT core, so you'll have to design that as well. You need a SNOT cube with studs in all six directions. Just a handful of connection points for each of the panels is usually sufficient.

DIGITAL BUILDING TIP

When you import an LDraw file from Bram's Sphere Generator into Studio, you'll notice that each of the six panels is considered a *submodel*. This is a collection of parts that have been grouped together and are treated as a unit. You can always break the submodel down into individual parts by right-clicking the submodel and selecting **Submodel ▸ Release** (selecting all the individual parts and clicking **Submodel ▸ Create** will re-create the sub-model), but it is also possible to edit the submodel in place by selecting **Submodel ▸ Edit**.

All six panels are identical, so once you've created an interlocking design for one instance of the submodel, you can click **Return to Main Model** and see that all six copies of the submodel have been automatically updated.

MODIFIED SPHERES

With some modifications, Lowell spheres can be integrated into larger LEGO models as domes and other architectural features. For the rounded dome of my Taj Mahal model, for example, I started with a Lowell sphere with a diameter of 27.2 studs. It consisted of a 16×16×16 SNOT core and six 14-plate-thick curved panels. To turn this sphere into a dome, I removed the bottom panel entirely and reduced the height of the SNOT core by 2 studs.

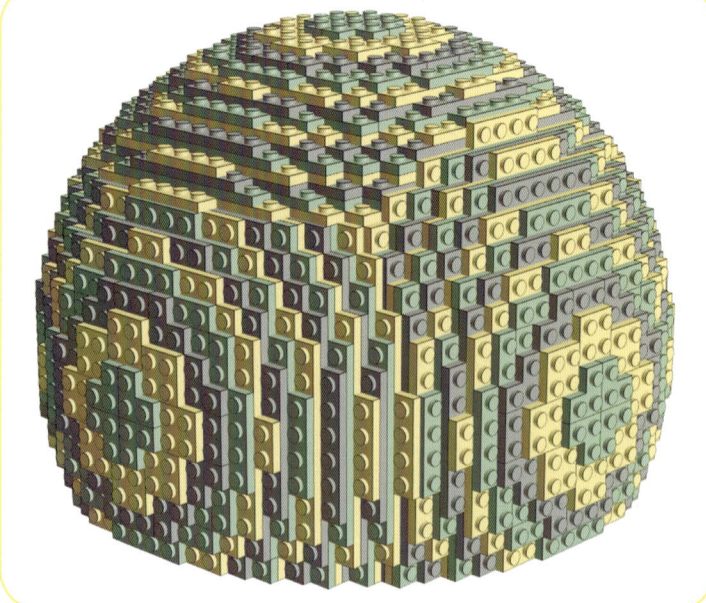

Figure 7-18: Truncating a Lowell sphere to form a dome

Then I cropped the bottom portions of the four side panels to create a flat bottom surface and planned out a fully interlocking design for each panel (Figure 7-18).

The panels on the front and back are oriented differently from the ones on the left and right. Truncating the bottom portions of all four at the same level yields two unique variants of the panels, one for the front and back and another for the left and right. Figure 7-19 shows the final dome of the Taj Mahal in the correct color scheme and with the decorative elements added on top, along with a breakdown of its various components.

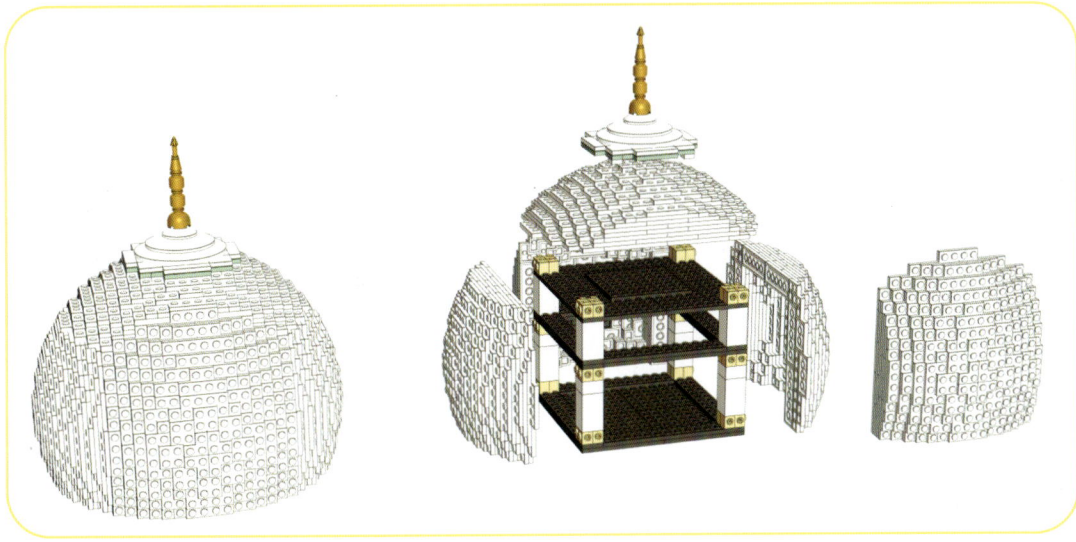

Figure 7-19: The dome of my Taj Mahal model

The bottom portion of the dome, which sits on the drum (shown in Figure 7-11), ended up having a diameter of 16 studs + 22 plates = 24.8 studs, very close to the drum's outer diameter of 25 studs.

SUMMARY

As you've learned in this chapter, you don't always need round bricks to create round LEGO shapes. It's possible to approximate round shapes with square (and rectangular) elements by using techniques like brick bending, hinged polygons, and Lowell spheres. This chapter also gave you a taste of how to use software tools like Bram's Sphere Generator to help plan LEGO builds. We'll continue to explore software-assisted building in the remainder of the book as we consider LEGO mosaics (Chapter 8) and sculptures (Chapter 9).

PART III

COMPUTER-ASSISTED BUILDS

Mosaics and sculptures aren't the kinds of LEGO builds you create just by playing around with a random pile of bricks. Rather, they require careful planning. To avoid the tedium of working out these builds manually, many people turn to software to automate the process. The following chapters outline some of the available tools and give you a better appreciation for what happens under the hood when you use software to design LEGO mosaics and sculptures.

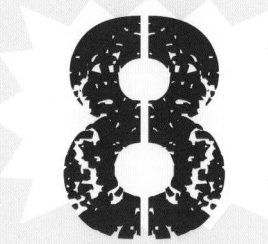

MOSAICS

Mosaics are a traditional art form dating back to ancient times, in which a picture or pattern is made from an arrangement of small colored tiles, stones, pieces of glass, or other materials. As a system of small, colored pieces, LEGO is a natural medium for creating mosaics by arranging bricks, plates, tiles, or other elements in a grid to form a 2D image. It's no wonder, then, that designing LEGO mosaics is a popular aspect of the hobby.

In this chapter, we'll consider the different types of LEGO mosaics and discuss the steps involved in making one. Often, computer programs are used to aid with the process of designing a LEGO mosaic. Understanding the techniques the programs employ can help you better leverage these tools to create realistic, eye-catching results.

FROM IMAGE TO MOSAIC

The typical starting point for most LEGO mosaics is a recognizable image such as a photograph or painting, the goal being to find a way to translate the contents of the image into a grid of different-colored LEGO elements. In the early days of the LEGO hobby, this translation process had to be worked out manually: the builder might divide the image into a grid and choose the best LEGO color to match each cell of the grid. The meticulous nature of this work greatly limited the size and complexity of the resulting mosaics. More recently, the advent of digital images and modern computers has made it possible for builders to turn any picture of their choosing into a sophisticated LEGO mosaic, using software tools to help automate the process.

To see how software can aid in the creation of a LEGO mosaic, it helps to understand how the data in a digital image file (say, a JPEG, PNG, or BMP file) is organized. An image is basically a 2D grid of square *pixels* (short for *picture elements*), each of which has a particular color (see Figure 8-1). When there are enough of them, the pixels blend together in our brains to create the illusion of a continuous image, with light, shadows, shapes, colors, and textures that resemble what we see in the real world.

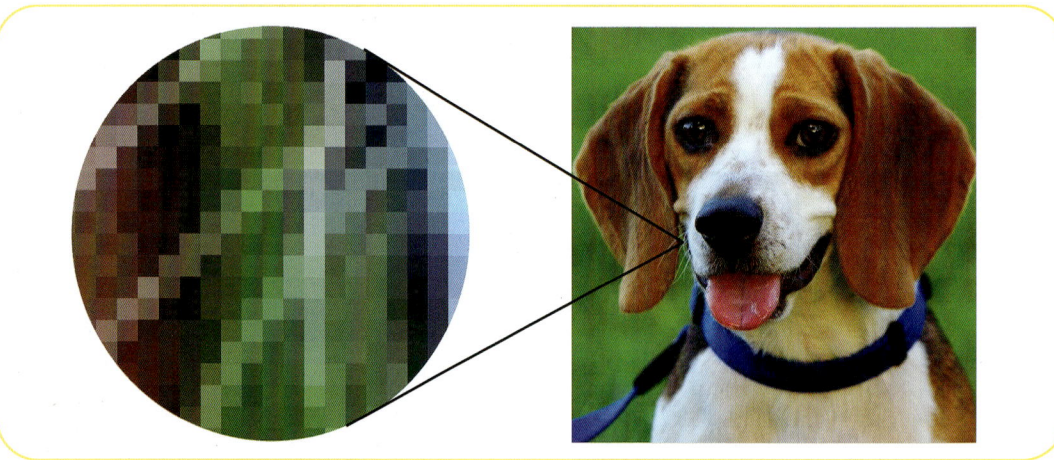

Figure 8-1: Zooming in on a digital image reveals how it's made up of square pixels, each with its own color.

If you think about it, a digital image is already a kind of mosaic—its tiny, colored pixels are equivalent to the small stones or tiles ancient artists organized into intricate designs. The pixel grid in a digital image is also analogous to the regular square grid of stud locations in the LEGO system. Software makes it easy to bridge the gap between the mediums, automatically examining the pixels in an image and selecting the best arrangement of LEGO colors to represent them.

It's tempting to think we could simply map each pixel in an image to an individual 1×1 LEGO brick or plate in the right color to create a LEGO mosaic. Unfortunately, it isn't as straightforward as that. In fact, a LEGO mosaic can at best be only a crude approximation of the original image. There are two reasons for this: the resolution of the image and its color depth.

RESOLUTION

The *resolution* is the level of detail captured in a digital image. It's proportional to the number of pixels the image contains: the more pixels the image has, the more detail it's able to capture. Digital images typically have a high enough resolution that you can't see the individual pixels that make up the image (unless you zoom in on it all the way using a photo editor like Photoshop). There are so many pixels that they all blur together. By contrast, you normally *can* see the individual tiles that make up a mosaic, unless you're quite far away, and so the level of detail captured in a mosaic is typically lower. In a nutshell, the size of an individual pixel relative to a whole digital image is much smaller than the size of an individual tile relative to a whole mosaic.

To put this into perspective, let's consider an example. The earliest digital cameras from the late 1990s produced images with something like 1,024×768 pixels. This would be considered low resolution today, when resolution is often measured in megapixels (a unit equivalent to 1 million pixels). Even if we tried to take one of these comparatively low-resolution 1990s images and translate it pixel-for-pixel into a LEGO mosaic, the result would be unwieldy. Given that the smallest element we can use to represent each pixel is a 1×1 brick or plate (with a footprint of 0.8×0.8 cm), we would end up with a LEGO mosaic that's a minimum of 27×20 feet!

Clearly, we have to dramatically reduce the resolution (number of pixels) of the source image before we can represent it using LEGO. For instance, if we downsized a 1,024×768 image by a factor of 8, we'd end up with 128×96 pixels. This would work out to a large (but much more manageable) LEGO mosaic that's 40×30 inches. However, the downsizing process comes with a significant loss of fine detail in the resulting LEGO mosaic. It's also more challenging to create the illusion of a continuous image because it's hard not to see the individual pixels that make up a LEGO mosaic. This is why most LEGO mosaics don't look very good when viewed up close. You have to step back a few feet to get the intended effect.

COLOR DEPTH

The *color depth* is the maximum number of colors that a digital image can represent. Once again, it helps to understand a bit about how image file formats work to see how an image's color depth affects the corresponding mosaic. Behind the scenes, an image file is nothing more than a whole bunch of numbers, each providing information about one of the image's pixels. In a typical color image, each pixel is represented using three 8-bit binary numbers, one for each of the three primary colors—red, green, and blue. By combining the three primary colors in the correct proportion, we can represent more colors in the color spectrum than humans can actually perceive. Each bit can have two values (0 or 1), and so 8 bits allow us to represent a total of $2^8 = 256$ values (0 to 255) for the intensity of each primary color. Put all three colors together and we have a total of $256 \times 256 \times 256 = 16.7$ million possible combinations, each of which represents a distinct color.

LEGO's color palette has expanded over the years, but the company hasn't yet produced bricks in 16.7 million different colors. So when we create a LEGO mosaic, we have to consider how to winnow the vast color depth of a standard digital image down to the 40 or so commonly available colors of LEGO elements. As with the resolution problem, this also results in a significant loss of detail. LEGO mosaics can't fully capture the rich gradations of light, shade, and color found in most digital images.

LEGO's narrower color palette also limits the kinds of images that can be represented well as mosaics. Grayscale images, or other images that hinge on many subtle variations of a small number of colors, would be difficult to translate to LEGO. By contrast, it's fitting that one of LEGO's first official mosaic sets geared toward AFOLs was a re-creation of Andy Warhol's famous screenprints of Marilyn Monroe. The pop art style of the original features a reduced color palette without subtle gradations, perfect for the low-resolution, low-color-depth nature of LEGO mosaics.

OFFICIAL LEGO MOSAICS

Although AFOLs have been building LEGO mosaics for a long time, the LEGO Group's first major foray into mosaics didn't come until 2020, when the company introduced a new theme called LEGO Art. Unlike the typical set, which contains a wide assortment of LEGO elements, the sets released under this theme mainly consist of a large quantity of 1×1 round plates or tiles in various colors. Once arranged according to the provided instructions, these 1×1 pieces come together to form 2D portraits of real and fictional icons, including Marilyn Monroe, the Beatles, Iron Man, and Darth Vader.

The LEGO Art sets also include 16×16 Technic bricks (#65803) for use as baseplates. While thicker than standard LEGO baseplates, these bricks can be easily joined together using Technic pins to create a large, flat surface for attaching the 1×1 elements.

TYPES OF LEGO MOSAICS

LEGO mosaics are generally classified based on the orientation of the LEGO pieces used to create them, studs-out and studs-up being the two main types. We'll look at these two types, along with lenticular mosaics, a variation on studs-out mosaics that combines two separate images into a single mosaic.

STUDS-OUT

Studs-out mosaics are the most common type of LEGO mosaic. They're created by attaching bricks or plates to a LEGO baseplate with their studs facing out, as shown in Figure 8-2. Each pixel from the downsized version of the original image is represented by a 1×1 brick or plate, which happens to have a square footprint, just like a digital pixel.

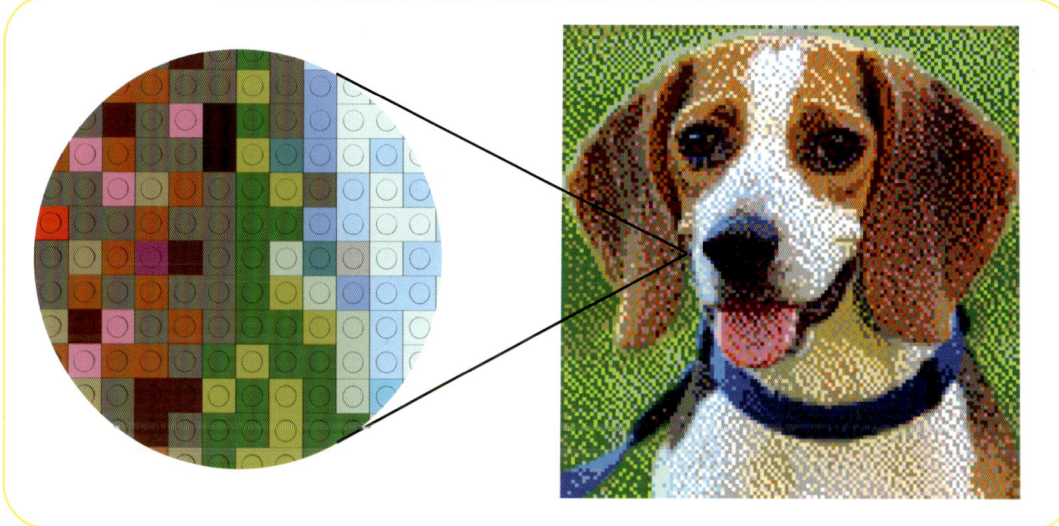

Figure 8-2: A studs-out mosaic

One advantage of using regular bricks or plates is that they can be combined into bigger pieces when the mosaic pattern allows it, reducing the overall part count. If two neighboring pixels have the same color, for example, you can use a single 1×2 element instead of two 1×1 elements. However, some builders like to avoid having blocky pixels in their studs-out mosaics, preferring instead to use 1×1 round plates or tiles (as in the LEGO Art sets) at the expense of a higher part count.

STUDS-UP

A *studs-up mosaic* is built from many adjacent columns of bricks or plates, stacked with their studs facing up. The design comes across by viewing the mosaic from the side rather than from the top, as in a studs-up mosaic. See Figure 8-3 for an example.

Figure 8-3: A studs-up mosaic can capture more detail in the same overall size.

When built using LEGO plates, a studs-up mosaic can achieve a higher resolution than the studs-out equivalent, packing more detail into the same overall size. This is because a plate is 2.5 times thinner than the regular stud dimension. The gain is only in the vertical dimension, however. Horizontally, each unit in a studs-up mosaic is still 1 stud wide.

A downside of the studs-up technique is that the "tiles" in the mosaic are rectangular, so they don't map perfectly to the square pixels in the original image. Practically speaking, studs-up mosaics are also more challenging to build because they're created by stacking layer after layer of plates rather than by attaching elements to a baseplate in a single layer.

The primary unit of a studs-up mosaic is typically the 1×1 plate, though the part count can be optimized by combining pieces with like colors into longer plates within the same layer. (You can also combine three-high stacks of plates with the same size and color into bricks, though some builders may prefer not to mix plates and bricks within a studs-up mosaic.) Even with these consolidations, you'll typically end up with many long columns of vertically aligned joints, which can lead to stability issues. The best solution is to make the mosaic 2 studs deep and alternate the orientation of plates between layers, much like the English bond pattern discussed in Chapter 3. In the odd layers, use plates that are 2 studs deep. In the even layers, use two rows of plates that are 1 stud deep—a front row maintaining the pattern of the mosaic, and a back row (hidden from view) of longer plates to strap everything together.

While studs-up mosaics are usually built using plates, some may prefer to use only bricks instead. This results in larger pixels that still aren't perfectly square (given that a 1×1 LEGO brick is slightly taller than it is wide). With bricks instead of plates, the resolution of a studs-up mosaic is actually slightly lower than a studs-out mosaic of the equivalent size.

LENTICULAR

The image in a *lenticular mosaic* changes with the viewing angle. This type of mosaic takes its inspiration from a 16th-century invention called *tabula scalata* (or "turning images"), in which two images are divided into vertical strips and printed on different faces of a folded surface, such that one image is visible from one viewing angle and the other image is visible from another angle.

In 2010, Chris Doyle had the idea of reproducing this effect in LEGO using 1×1 cheese slope pieces (#54200). He placed the cheese slopes such that the slopes faced opposite directions (left versus right) in alternating columns (see Figure 8-4). This created a zigzag, folded surface, much like the traditional tabula scalata, and allowed two distinct images to be combined into a single LEGO mosaic—one from the colors of the left-facing cheese slopes and the other from the colors of the right-facing cheese slopes.

Figure 8-4: A close-up on a section of a lenticular mosaic

Figure 8-5 shows an example of a lenticular mosaic with two different images that can be seen depending on the angle that the mosaic is viewed from.

Figure 8-5: A lenticular mosaic, viewed from both angles

Lenticular mosaics can be built much like studs-out mosaics, with cheese slope pieces attached to one or more baseplates or 16×16 Technic bricks. If two adjacent 1×1 cheese slope pieces in the same column have the same color, it's possible to replace them with a single 1×2 slope piece (#85984) to reduce the part count.

OTHER TYPES OF MOSAICS

There are several other kinds of LEGO mosaics created using one or more specific types of LEGO pieces. These usually represent abstract patterns (geometric or otherwise) instead of more detailed images such as paintings or photographs. They're also typically created by hand instead of using software (except for 3D mosaics, which may use a 2D design created with software as a starting point).

THREE-DIMENSIONAL

A *3D mosaic* aims to add depth to the picture or pattern depicted in the mosaic. This is achieved by stacking multiple layers of bricks or plates in certain areas of the mosaic. Figure 8-6 shows an example, created by Dana Meyrow. The 3D effect reinforces the wave pattern created using the various shades of blue.

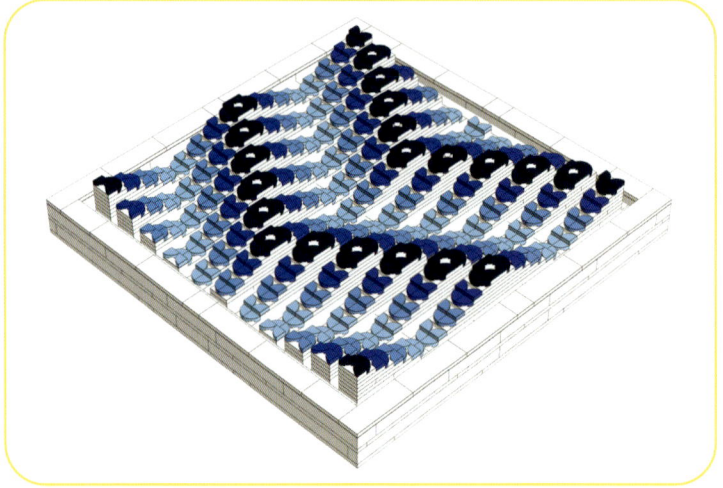

Figure 8-6: A mosaic with plates stacked to different heights to create depth

It's also possible to create a mosaic using a grid of more unusually shaped LEGO elements instead of regular bricks or plates, such as the minifigure mosaic shown in Figure 8-7.

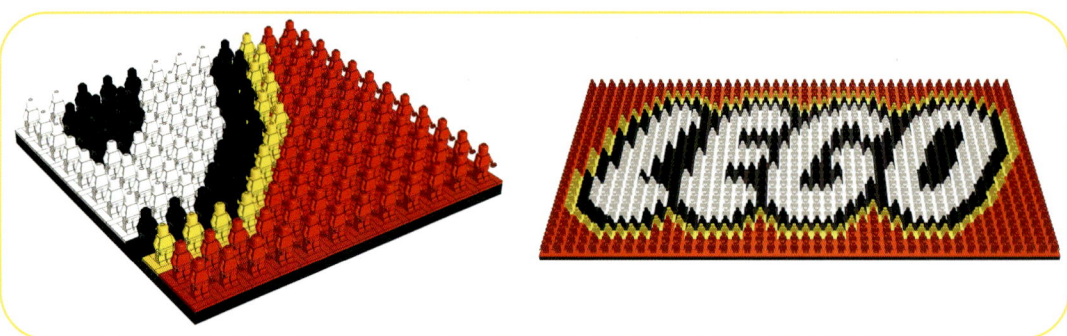

Figure 8-7: A 3D mosaic created using minifigures

Some LEGO artists even create mosaics that depict a familiar image when viewed from a distance but, upon closer inspection, turn out to be built from a seemingly random mishmash of LEGO elements, including unusual ones like wheels, gears, and so on. For example, the mosaic in Figure 8-8, created by Gerardo Pontiérr, depicts Spanish surrealist artist Salvador Dalí.

The variety of elements involved adds interesting surface texture to the mosaic. Despite the exotically shaped pieces, however, notice that the demarcations between one color and another still largely conform to the regular LEGO grid.

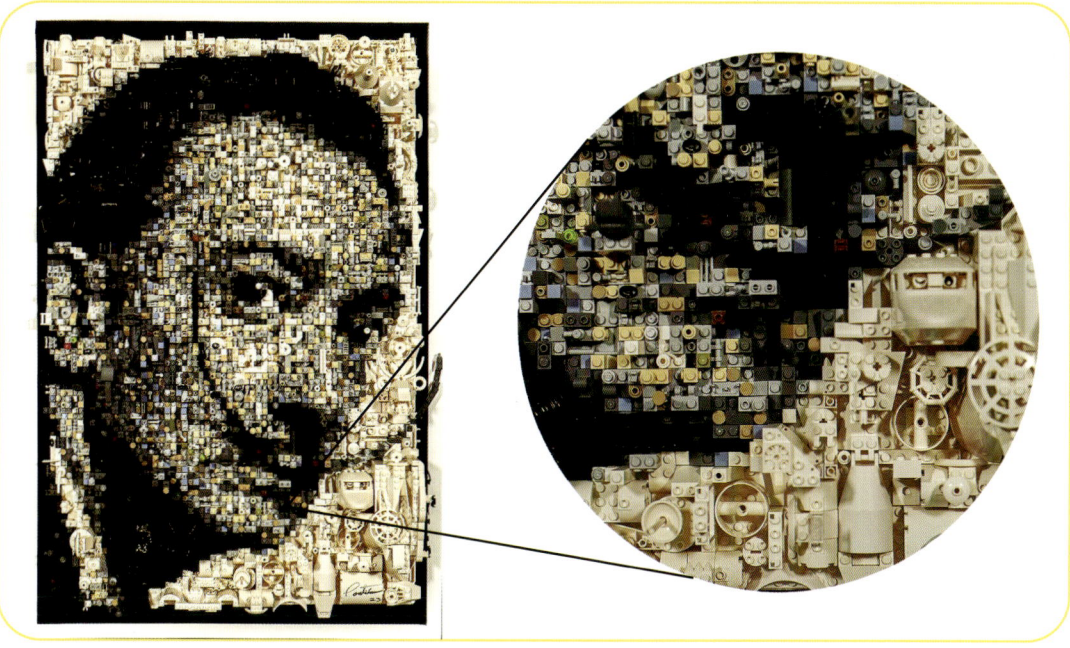

Figure 8-8: A mosaic of Salvador Dalí made from random LEGO pieces

ISOMETRIC

Isometric patterns are designs that don't use the traditional perspective distortion you'd normally see in a 2D representation of a 3D space. Instead of parallel lines converging on a vanishing point, the parallel lines remain parallel, which can lead to interesting optical illusions. Figure 8-9 shows an example LEGO mosaic of an isometric pattern.

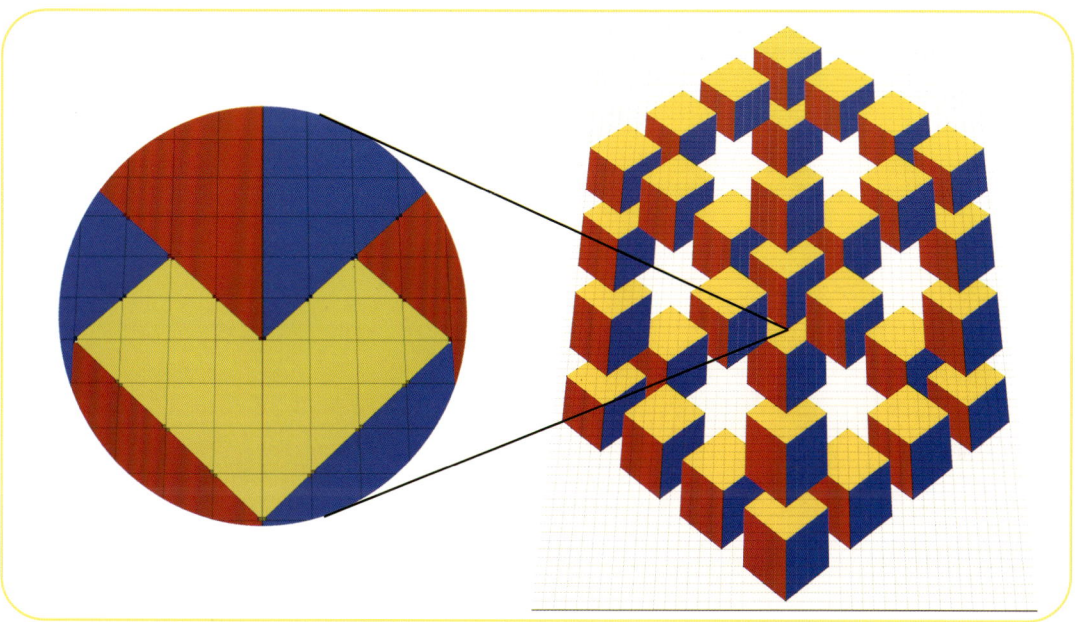

Figure 8-9: An isometric pattern mosaic

Isometric mosaics became possible with the advent of the 2×2 triangular tile (#35787) in 2018. These elements (used in combination with regular 2×2 square tiles) make it easy to create sharp diagonal edges between colors at a consistent angle. The result is an illusion of 3D forms built from flat, different-colored planes.

CHEESE SLOPE

Cheese slopes aren't just for creating lenticular mosaics; the unique shape of these elements allows for the creation of interesting geometric patterns when they're laid on their side, often in combination with 1×1 tiles. This style of LEGO mosaics was popularized by builders like Katie Walker. Figure 8-10 shows an example.

One tricky thing about cheese slope mosaics is that there are no stud connections inside the mosaic, so the pieces aren't really attached to anything. Instead, they're just pushed together tightly and held in place by an outer frame of bricks or plates. This makes mosaics built with this technique very fragile.

Figure 8-10: A cheese slope mosaic by Katie Walker

HEADLIGHT BRICK

Headlight bricks, as discussed in Chapter 5, can be joined together in quite a few different ways. Headlight brick mosaics exploit this flexibility to create interesting abstract patterns. Seen in Figure 8-11 is one such pattern, inspired by the work of Brendan Powell Smith.

142

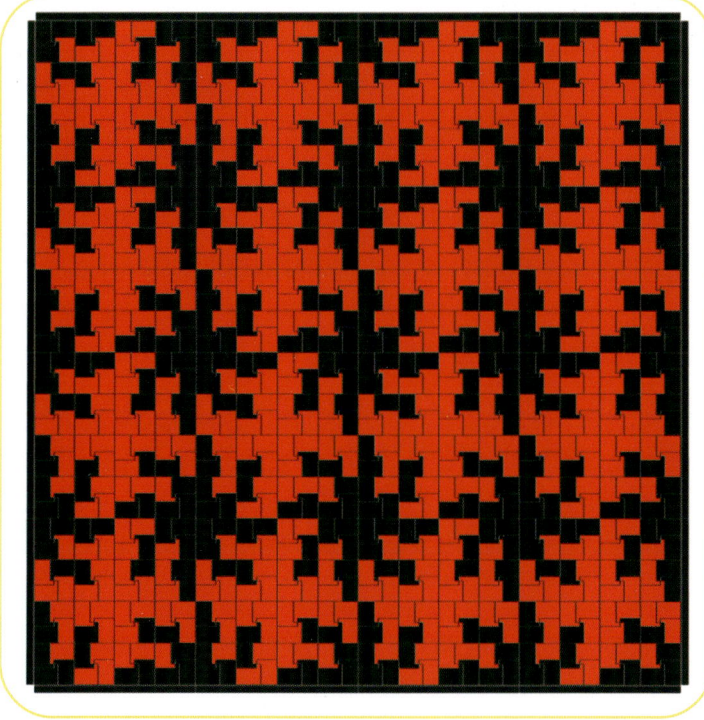

Figure 8-11: A headlight brick mosaic

Headlight brick mosaics have the advantage that they are fully connected and don't require a baseplate to hold the pieces together. Also, there's a seemingly endless variety of patterns that can be created using just this one type of LEGO element, especially if you use two or more colors.

MOSAIC DESIGN SOFTWARE

Numerous online software tools are available to help automate the process of creating LEGO mosaics from your own images. Some of these must be downloaded and installed on your computer, while others offer convenient web-based interfaces for uploading your images and generating mosaics that can be downloaded almost instantly (as LDraw files or PDFs containing instructions for assembling the mosaic). Under the hood, all these programs follow the same basic steps to turn the provided image into a LEGO mosaic. They differ in the amount of control they offer you for tweaking parameters to influence the final design.

BrickLink Studio includes a mosaic tool, though its capabilities are somewhat limited. You can upload an image of your choosing and make basic adjustments (crop level, brightness, contrast, saturation, and so on). You then pick the size of the LEGO mosaic you want to create as well as the color palette (a subset of the LEGO color palette) and the types of pieces (square bricks, plates, tiles, or 1×1 round plates or tiles) to use for your studs-out mosaic. There's also an option to optimize your part count by combining adjacent pieces of like colors into bigger pieces (assuming you aren't using round elements).

An option that's far more powerful is LEGO Art Remix (*https://lego-art-remix.com*), a web-based program developed by Deb Banerji. LEGO Art Remix is easy to use and includes a wider array of options for controlling each step in the process of converting your image into a LEGO mosaic. In the coming sections, we'll walk through these steps and some of the advanced settings you may encounter in LEGO Art Remix or other mosaic tools. Not every program gives you visibility into the algorithms it's using to carry out the various steps, let alone the ability to choose which algorithms to use, as LEGO Art Remix does. Understanding the steps the program goes through behind the scenes can help you make more informed decisions when designing your own mosaics.

STEP 1: UPLOADING THE IMAGE AND SETTING THE SIZE

The first step in creating a mosaic is to choose an image and upload it into the software you're using. Of all the steps in the process, this one perhaps requires the most thought, as not every image will make a good LEGO mosaic. For example, images that are too dark or crammed with too much fine detail wouldn't be good candidates. Even images that work well may need to be fine-tuned in step 3.

Once you upload an image, you need to pick a size (in terms of studs) for the mosaic. This will affect the part count and the cost to build the mosaic using real pieces. You should also be mindful of how the stud dimensions translate into real-world dimensions. Remember, 1 stud is equal to 0.8 cm.

STEP 2: SELECTING THE PALETTE

Typically, the next step is to pick the palette of LEGO pieces that will be used to create the mosaic—both the color palette and the type of pieces. You can use any of the basic LEGO elements (bricks, plates, or tiles). You can also choose round elements or square ones, but remember that round 1×1 pieces can't be consolidated into larger elements to reduce the part count.

Some programs give you the ability to use checkboxes to select the colors you want to use out of the LEGO palette, allowing you to exclude colors you don't have or that may be more expensive or harder to find. LEGO Art Remix also has a great feature that lets you limit the mosaic to pieces found in one or more official LEGO Art mosaic sets. That way, if you already have the official sets, there's no need to order any extra pieces to build your mosaic.

STEP 3: PREPARING THE IMAGE

The next step is to prepare the image, optimizing it for conversion into a LEGO mosaic. Some of the basic settings include brightness, contrast, and saturation, and the software you're using should have slider controls that allow you to adjust them to your liking. It may help to turn up the brightness on a darker image, increase the contrast if the image has subtle color gradations, or punch up the saturation a bit to achieve optimal results.

Another important part of this step is resizing your image so its dimensions in pixels match your desired mosaic dimensions in studs. If the aspect ratio of the image is different from that of your mosaic, the image will also have to be cropped so the resizing process doesn't distort the image. The software will usually have controls that allow you to set the cropping area using your mouse.

The actual resizing happens under the hood of the software, but it's instructive to see how the resizing may work for the different types of mosaics we've covered. Say you want your LEGO mosaic to be 128 studs wide by 128 studs tall (which makes it about 40 inches on a side). The image will have to be resized to be 128×128 pixels—at least in the case of a studs-out mosaic.

The resizing is a little trickier for studs-up and lenticular mosaics. The pixels in studs-up mosaics are rectangular, and so we can fit more of them (2.5 to be exact) in 1 stud dimension vertically. The image will have to be resized to 128 pixels wide by 320 pixels tall, since 128 × 2.5 = 320. This is one case where the aspect ratio of the image deliberately doesn't match the aspect ratio of the final mosaic, but when you build the mosaic using the sides of 1×1 plates (which are much wider than they are tall), it will end up with the correct proportions.

For lenticular mosaics, two different images need to be combined within the overall width of the mosaic. Each image needs to be resized to be half the width but all the height (in terms of pixels) of the finished mosaic, which works out to 64 pixels wide by 128 pixels tall in our example. Once the columns from the two images are interleaved, the mosaic will have a total size of 128×128 studs.

STEP 4: QUANTIZING THE COLORS

The next step, *color quantization*, is the process of decreasing the color depth of an image—that is, reducing the number of distinct colors it uses. This is an important step for the creation of a LEGO mosaic: it gets you from an image that has potentially millions of distinct colors to one with just the 40 or so colors in the LEGO palette. The method used for this step has a great bearing on how closely the LEGO mosaic resembles the original image. We'll consider two concepts related to color quantization that you may encounter when you use mosaic tools like LEGO Art Remix: color distance and dithering.

COLOR DISTANCE

Color distance is a measure of how good a match one color is for another during the quantization process. We discussed that digital image files typically treat pixels as three values, each from 0 to 255, representing the levels of the three primary colors (red, green, and blue). Imagine using those three values as coordinates for a point in three-dimensional (3D) space. The available colors in the LEGO palette can also be expressed as RGB values and thought of as points in the same RGB color space. By thinking of colors spatially in this way, we can examine the distance between them to find the best match.

The easiest (and least computationally intensive) way to figure out the distance between any two colors in the RGB color space is to calculate the *Euclidean distance*, the length of the straight line joining the two points in the RGB space. Basic quantization looks at the color of each pixel in the original image and finds the spatially nearest neighbor (based on the Euclidean distance) of the colors in the more limited palette. This is the method most mosaic programs use.

Unfortunately, the RGB method doesn't always align with the way humans perceive similarity between colors. Programs like LEGO Art Remix give you the option to use other, more computationally intensive techniques that can result in better color accuracy in your mosaic. This involves mapping all the colors from the RGB space to an alternate color space known as the *Lab color space*, which was designed to better represent the way humans perceive color. Lab is a 3D space, where *L*, or *lightness*, is an achromatic component that represents shades of gray, and the *a* and *b* components represent the proportions of four primary colors (red, green, blue, and yellow).

Even within the Lab space, there are different ways to find the closest color match. LEGO Art Remix lets you use either the Euclidean distance within the Lab space, which is less accurate but quicker, or the Delta-E function (listed as CIEDE2000), which is the most accurate way of calculating the distance between two colors in the Lab space. Try switching between these different options and see how the results vary.

DITHERING

Another option that can be enabled in LEGO Art Remix is a technique called *dithering*, which can somewhat alleviate the loss of color depth from color quantization. Dithering involves grouping different-colored pixels in such a way that they collectively emulate other colors. With this technique, you can use a limited color palette—in this case, the LEGO color palette—to represent a wider gamut of colors. LEGO Art Remix offers a variety of dithering algorithms for you to experiment with. We won't focus on the nuances here, but rather get a high-level overview of how dithering works and how it can help create better LEGO mosaics.

Historically, dithering was used in early computer displays that had very limited color palettes. To this day, inkjet printers use *halftoning*, a form of dithering in which single-color dots of varying sizes and spacings simulate a smooth gradient of colors. We can apply this same technique to LEGO. For example, Figure 8-12 shows five different patterns created by arranging four 1×1 plates in two possible colors: black and white. The proportion of black increases as we move from the left to the right.

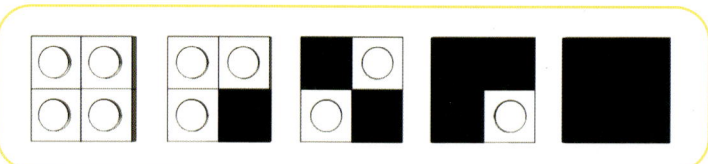

Figure 8-12: Five patterns with increasing proportions of black (from left to right)

Now consider a large mosaic (Figure 8-13) created by tiling these five patterns in five separate strips. Notice that the three strips in the middle appear to have increasingly darker shades of gray—at least when viewed from a distance. Squinting your eyes a bit may also help.

This is a simple illustration of how two basic colors (black and white) can be mixed in different proportions to create the illusion of a wider range of colors. In the same way, if the original mosaic image includes colors that don't exist in the LEGO palette, dithering can arrange LEGO pieces in the available colors in such a way that they combine in the viewer's brain to create the illusion of seeing the missing colors. Figure 8-14 shows two mosaics based on the same image: one created without and one with dithering.

Figure 8-13: A mosaic created by repeating the five patterns. The three patterns in the middle simulate different shades of gray.

146

Figure 8-14: Mosaics based on the same image without (left) and with (right) dithering

Notice how dithering in the version on the right greatly reduces the blotchiness resulting from the limited LEGO color palette in the version on the left. Dithering can therefore help represent subtle color gradations (especially skin colors in mosaics that show human faces) more accurately. Some people, however, may find the extra "grain" that dithering adds to be more bothersome than the blotchiness that occurs without dithering. Therefore, deciding to use dithering is a matter of personal preference.

STEP 5: CREATING THE ACTUAL MOSAIC

Once you've resized the image to match the dimensions of the mosaic and mapped its colors to the LEGO color palette, the last step is to create the actual mosaic by digitally laying out the necessary LEGO pieces. If the tool allows it (and you've enabled this option), part optimization may also occur at this stage. This is where adjacent pieces of like colors are combined into bigger pieces, reducing the overall cost of your mosaic (and hopefully making it a little less tedious to build). The output from this step can be a Studio file (BrickLink Studio), an LDraw file, or a list of pieces needed to build your mosaic in an XML format, along with the instructions in a PDF (LEGO Art Remix produces the XML and PDF).

SUMMARY

In this chapter, we reviewed the different types of LEGO mosaics and discussed how to use software tools such as BrickLink Studio or LEGO Art Remix to assist in the mosaic-design process. We've outlined the steps these tools use to convert an image file into a plan for a mosaic that can be built using real pieces. In Chapter 9, we'll look at another category of LEGO builds that benefit from the use of computer software: LEGO sculptures.

SCULPTURES

In the last chapter, we discussed how to create a 2D LEGO mosaic based on any image. But why confine ourselves to two dimensions when LEGO is a 3D medium? In theory, it should also be possible to use LEGO pieces to build a sculpture of any 3D object. In this chapter, we'll explore techniques for sculpting interesting shapes with LEGO.

WHAT IS A LEGO SCULPTURE?

A sculpture in a traditional sense is a 3D piece of art created using wood, clay, metal, or stone. Of course, we've been building in three dimensions throughout this book, creating models of houses and other structures. We've explored techniques for building straight walls, angled walls, and even round walls. How do LEGO sculptures differ from the builds we've already covered?

When we talk about *LEGO sculptures*, we're referring to models with shapes that can't easily be broken down into basic geometric components like cubes, prisms, cylinders, and spheres. It's impossible to define these shapes using a simple set of measurements. For instance, it's a far cry to go from building a LEGO model of your house to building a model of your dog (see Figure 9-1). Where would you even begin?

Figure 9-1: A LEGO sculpture of a dog

To design a model of something like a house, you'd typically start by getting the measurements and determining the right scale and proportions to use. But this doesn't seem very practical for something like a dog. There would be far too many measurements to keep track of (assuming the dog can sit still long enough for you to take the measurements in the first place). Also, unlike a typical house, a dog is made up of complex organic shapes. There's no good way to intuitively figure out how to arrange LEGO pieces to mimic those shapes. As with creating mosaics, this is an area where software can come to the rescue.

WORKING WITH 3D MODELS

If there were a 3D equivalent of an image file that contained information about the shape of an object, we could use software to convert that information into a LEGO model. In fact, such files do exist. They're called *3D models*, and they come in formats such as STL, Wavefront OBJ, PLY, and Collada. These files are primarily used for 3D printing and computer graphics applications, but we can also harness them to help create LEGO sculptures. The vast majority of the LEGO sculptures you've seen online or at a LEGOLAND were probably based on 3D models.

You can create your own 3D models using programs like Blender, ZBrush, Cinema 4D, and the like, but that process is beyond the scope of this book. Fortunately, there's a wide selection of 3D model files available online, either for free (on sites like *https://www.thingiverse.com*) or for a nominal cost. You can probably find a model of just about anything you'd want to build, from humans, animals, and different modes of transportation (cars, ships, airplanes, and so on) to iconic real-world sculptures.

UNDERSTANDING STL FILES

To understand how a 3D model file can form the basis of a LEGO sculpture, it helps to know a bit about how these files are organized. The details vary from one file format to another, but we'll consider the most common format, STL, as a representative example. This format takes its name from the word *stereolithography*, a type of 3D printing. However, people assume that STL stands for Standard Tessellation Language or Standard Triangle Language because tessellation and triangles are two keywords related to how STL files are able to describe 3D surfaces.

Tessellation is the process of covering a surface with a network of one or more types of geometric shapes such that there are no gaps or overlaps between shapes. Essentially, this is the 3D equivalent of covering a floor with tiles. Different shapes can work, but STL files specifically use triangles. In fact, any complex 3D surface can be tessellated using only triangles (see Figure 9-2). The resulting network of interconnected triangles is known as a *mesh*, and each individual triangular surface is called a *facet*.

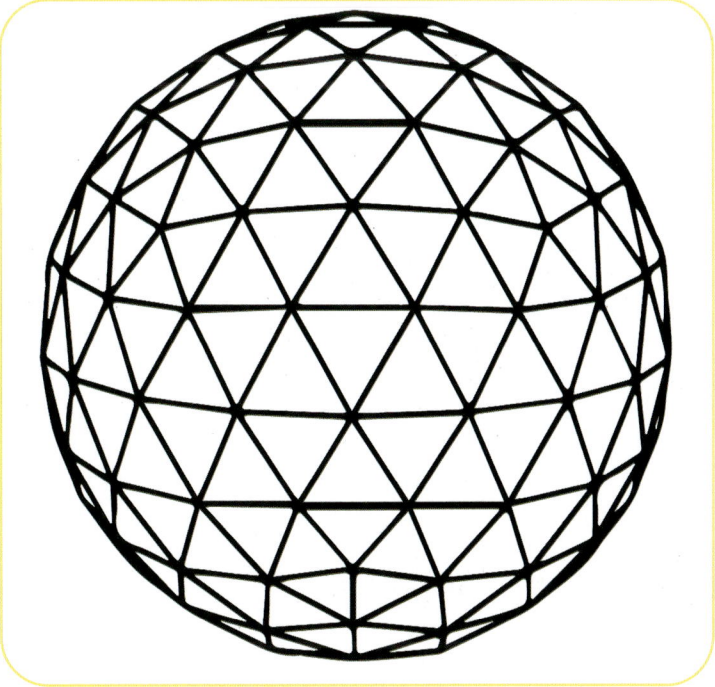

Figure 9-2: The surface of a sphere tessellated with triangles

151

Each triangle in the mesh has three sides and three corners, or vertices. An STL file primarily contains a collection of x, y, and z coordinates describing the location of every triangle's vertices in 3D space. The file also includes a *surface normal* for each facet, indicating which face points outward. Surface normals are used in computer graphics applications to create realistic lighting effects; they help determine how light is reflected off the surface of the 3D model. STL files can be in a binary format (for reduced file sizes) or a regular text format. If you were to open one of the files in a text format, you would see lots of entries like the following, one for each of the triangles that make up the mesh:

```
facet normal  0.146978E+00  0.437754E-01 -0.988171E+00
    outer loop
        vertex  0.600648E+02  0.582387E+02  0.589680E+01
        vertex  0.610294E+02  0.550000E+02  0.589680E+01
        vertex  0.550000E+02  0.550000E+02  0.500000E+01
    endloop
endfacet
```

Each vertex line gives the 3D coordinates for one of the triangle's corners.

A simple 3D shape like a cube can be represented using as few as 12 triangles, two for each face of the cube (Figure 9-3, left). Meanwhile, a more complex organic shape requires many more triangles to accurately represent its surface geometry. The heart on the right of Figure 9-3 consists of around 6,000 triangles, for example. In general, the more triangles in the mesh, the more accurate the 3D model will be, but the more computationally intensive the model will be to work with.

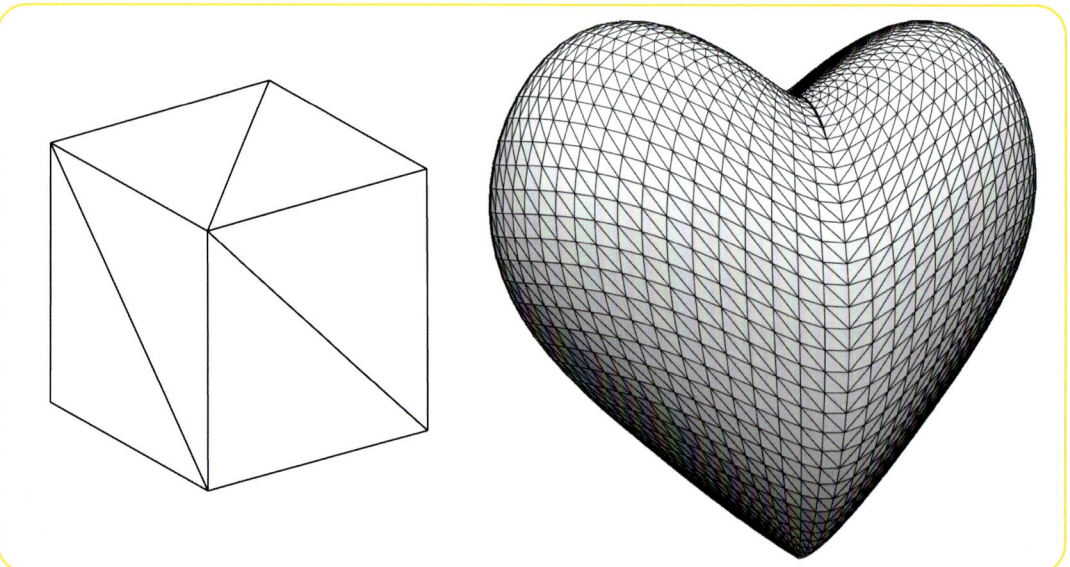

Figure 9-3: A cube can be represented using 12 triangles, whereas a more complex, rounded shape like a heart requires many more.

152

MANIPULATING A MODEL

Once an STL file is loaded into a program that can work with it, it's possible to manipulate the model in various ways, such as:

TRANSLATION   Shifting the model's position in 3D space

ROTATION   Turning the model about one or more of its axes

SCALING   Expanding or shrinking the model in one or more dimensions

As we'll see a little later, scaling is an especially important step in the process of creating a LEGO sculpture. It allows us to compensate for the geometry of the pieces that will make up the LEGO equivalent of the 3D model.

TYPES OF LEGO SCULPTURES

Just like mosaics, LEGO sculptures are classified based on the orientation of the LEGO pieces used to create them. Let's consider the differences between the two main types of sculptures: studs-up and studs-out.

STUDS-UP

Studs-up sculptures are created by stacking bricks in a conventional studs-up way. This is the most common type of LEGO sculpture, especially since it's the easiest kind to create. Figure 9-4 shows a studs-up sculpture created using the 3D model of the heart that we saw in Figure 9-3.

Figure 9-4: A studs-up sculpture of a heart

A studs-up sculpture can only be a coarse approximation of the original 3D form, especially if there are curves involved. The jagged "step" from a brick in one layer to a brick in the next just isn't the same as a smooth curve. However, you can achieve a higher "resolution" for the sculpture by mixing in some LEGO plates along with regular bricks. After all, plates are one-third as tall as bricks, so they provide three times as many gradations of height, allowing you to capture more detail in the same overall size. For example, Figure 9-5 shows a revised heart sculpture with some plates mixed in.

Figure 9-5: A studs-up heart sculpture with mixed plates and bricks

As we saw with spheres in Chapter 7, adding plates only helps refine the shape in the vertical (z) dimension, since plates have the same footprint as bricks in the other two dimensions. In other words, only the surfaces of the sculpture that face upward and downward can benefit from the use of plates, while no extra detail is added to the sides of the sculpture. But even in this one dimension, the addition of plates can be worth it. In Figure 9-5, for example, notice that the curved effect on the top surface of the heart sculpture is smoother thanks to the extra gradations of height.

STUDS-OUT

Studs-out sculptures extend the idea of the Lowell sphere (see Chapter 7) to sculptures in general. If plates help preserve more fine detail than is possible using bricks, but only in the direction that their studs are facing, why not use SNOT techniques to build the sculpture with its entire outer surface lined with plates? These plates could have their studs facing out in one of six possible directions (top, bottom, front, back, left, or right). For example, Figure 9-6 shows a studs-out version of the heart sculpture. Notice that it's a lot less blocky than the studs-up versions, with gradations of plates on all six faces creating a smoother, more curved appearance.

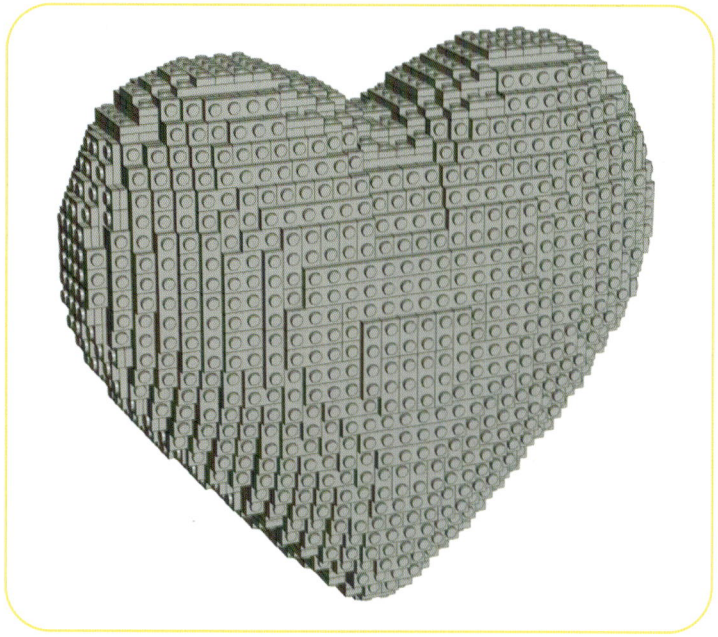

Figure 9-6: A studs-out sculpture of the heart shape

Studs-out sculptures can be challenging to design and build. How exactly do you determine the orientation of each plate in the sculpture? How do you ensure that the plates pointing in the six different directions mesh together, with no overlaps or gaps? Lastly, unlike a studs-up sculpture, a studs-out sculpture can't support itself. It needs a SNOT core on the inside to support it (much like a Lowell sphere). The plates facing the six different directions would form six panels that can be attached to the SNOT core. But unlike the case of the Lowell sphere, the six panels wouldn't be identical and may each be composed of multiple disconnected sections.

DESIGNING LEGO SCULPTURES WITH SOFTWARE

Unlike LEGO mosaics, the software options for generating LEGO sculptures are rather limited. Of the options that are available, the sculpture feature in BrickLink Studio is the easiest to use. You can import 3D models in various formats into Studio and create basic studs-up LEGO sculptures using bricks. By default, each sculpture will be composed entirely of 1×1 bricks, but this wouldn't be practical to build in real life, so there's also an option to use longer pieces wherever possible to create an interlocked structure.

Another option is Bricker (*https://www.blendermarket.com/products/bricker*), an add-on for the Blender 3D modeling program, but unless you're already savvy with Blender, it comes with a steeper learning curve. Created by Christopher Gearhart, Bricker allows you to create studs-up LEGO sculptures from 3D models, with many more customization options at your disposal than you'll find in Studio.

Last but not least is LSculpt (*https://lego.bldesign.org/LSculpt*), currently the only option available for creating studs-out sculptures. It was developed in 2006 by Bram Lambrecht (who had earlier created the program we discussed in Chapter 7 for generating Lowell spheres of any size). LSculpt generates the outer shell of the studs-out sculpture entirely from 1×1 plates. It's then up to the user to do the manual work of combining those 1×1 plates into interlocked panels and creating the SNOT core to hold them together. There's no fully automated solution available for generating structurally sound studs-out sculptures.

SOFTWARE-ASSISTED STUDS-UP SCULPTURES

The following sections outline the basic steps involved in creating a studs-up LEGO sculpture with software assistance. (We'll turn to LSculpt and studs-out sculptures later in the chapter.) Some of these steps, especially steps 3 and 4, happen behind the scenes in the software tool you're using, but they're directly affected by the input you provide in steps 1 and 2. Understanding what the software is doing can help you design LEGO sculptures more effectively.

COLOR IN 3D MODELS

The STL format doesn't natively support information about color. It was developed during the earliest days of 3D printing, when color wasn't a consideration, and it hasn't changed since. Some of the other 3D modeling formats support colors, either by including color information for each vertex or facet in the mesh or by containing separate material or texture information. Certain LEGO tools, like Bricker, can leverage this color information and map to the closest LEGO colors when the 3D model is converted into a LEGO sculpture. In this chapter, we'll limit our discussion to 3D models that are devoid of color information and leave adding color to the manual process of actually realizing the LEGO sculpture.

STEP 1: UPLOADING AND SIZING THE 3D MODEL

The first step in creating a LEGO sculpture is to upload the corresponding 3D model to your chosen software tool. At this stage, you also specify your desired size of the sculpture in terms of studs along a particular axis—say, the x-axis. The tool then automatically determines the size of the sculpture in the other two dimensions using the proportions of the 3D model.

Programs can sometimes have trouble working with 3D models that have a large number of facets (which can run into the millions for complex shapes). There's a process known as *decimation* that can simplify your 3D model before you upload it, and it's available in most 3D modeling tools (like 3D Builder in Windows). Decimation combines the triangles in a mesh to reduce their total count. This creates a coarser but computationally lighter representation of the original shape.

STEP 2: SPECIFYING THE ORIENTATION

The different programs that create 3D models don't always agree on which axis points up. Once you upload your 3D model, you may need to reorient it such that your studs-up sculpture has LEGO pieces in your preferred orientation. BrickLink Studio, for instance, allows you to choose which direction in the model should be pointing up in the final sculpture.

STEP 3: PREPARING THE 3D MODEL

Under the hood of your software, the next step is to prepare the 3D model for translation into LEGO bricks. The model may need to be rotated along one or more axes depending on if and how you choose to reorient the model in step 2, for example. Also, depending on whether you're using bricks or plates, the 3D model will have to be scaled to compensate for the fact that LEGO pieces aren't perfect cubes.

STEP 4: CONVERTING TO VOXELS

The heart of the model-to-sculpture conversion process is *voxelization*, in which the 3D model is broken down into a 3D grid of cubes known as *voxels*. For example, Figure 9-7 shows a voxelized version of the heart shape from Figure 9-2. You can think of the voxels here as the 3D equivalent of the regular square grid of pixels that make up a 2D image.

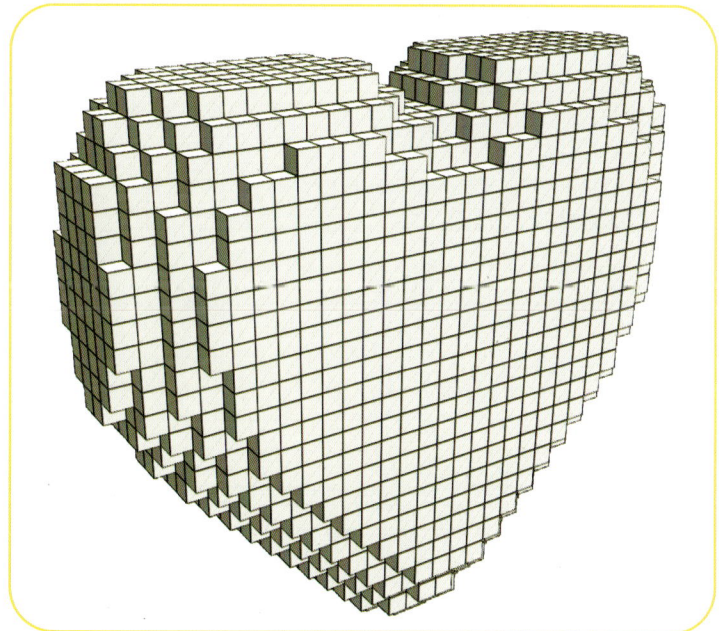

Figure 9-7: The heart shape broken up into voxels

The simplest voxelization algorithms work by subdividing each triangle in the original 3D mesh into smaller triangles until each side of each triangle is smaller than the size of each voxel (known as the *voxel pitch*). Then each voxel location in the grid that contains one or more triangles inside it is marked as occupied.

But what should the voxel pitch be? The software determines it based on the input you've provided—usually the desired number of studs in one of the three dimensions. For instance, if you're looking to build a sculpture that's 16 studs wide along the x-axis, the tool would internally set the voxel size such that the voxelized representation of the 3D mesh has 16 voxels along the x-axis.

You might think that creating a studs-up sculpture with bricks is just a matter of voxelizing a 3D model and replacing each voxel with a 1×1 LEGO brick. While this is essentially what happens at a conceptual level, in practice it isn't that simple, since unlike voxels, LEGO bricks aren't perfect cubes. As we've discussed, they have a height-to-width ratio of 6:5, so simply placing a LEGO brick in each occupied voxel location would result in a vertically stretched sculpture. To get the right proportions, the software has to scale the vertical (z) axis of the 3D model by a factor of 5/6 in step 3, *before* voxelization. That way, once the mesh is voxelized and each voxel is replaced with a 1×1 brick, the correct proportions are restored in the resulting sculpture. This is just like the process we went through in Chapter 7 to generate a studs-up sphere.

In the case of a studs-up sculpture that incorporates plates, it's best to think of each voxel as representing a 1×1 plate rather than a 1×1 brick. For that, the original 3D model must first be scaled by a factor of 5/2 in the vertical (z) axis. This compensates for the 2:5 height-to-width ratio of a 1×1 plate.

STEP 5: CREATING THE ACTUAL SCULPTURE

Even once the model is scaled so that its voxelized representation accounts for the shape of the LEGO pieces, realizing a sculpture isn't as simple as replacing each voxel with a LEGO piece. For instance, it wouldn't be sturdy to create a sculpture simply by stacking 1×1 bricks. Instead, the individual voxels need to be consolidated into longer bricks and placed in a way that creates a fully interlocked structure. Determining the best way to do this is a complex mathematical problem that's the subject of many scientific papers. Most software that creates LEGO sculptures addresses the problem by using the basic techniques we covered in Chapter 3: overlapping bricks to avoid joints that line up either vertically or horizontally, alternating the orientation of bricks between layers, and so on. Additionally, large studs-up sculptures are usually built hollow with some kind of internal framework (that the software may or may not generate) to provide extra support. This framework can use either regular bricks or Technic beams. Large sculptures installed in public spaces may also have additional internal reinforcements such as a framework made of steel or aluminum.

If the sculpture uses plates, it is best to consolidate three-high stacks of plates into bricks wherever possible. This results in a sculpture that's made up of bricks for the most part, with some plates used on the top and bottom surfaces for improved detail. Otherwise, building a studs-up sculpture entirely with plates would result in a needlessly high part count.

SOFTWARE-ASSISTED STUDS-OUT SCULPTURES

The process involved in creating a studs-out sculpture is more involved than the steps we just outlined. As mentioned, LSculpt is currently the only program that can generate studs-out sculptures. It hinges on the fact that stacking 20 1×1 plates in four columns of 5 plates each creates a perfect cube (if you disregard the studs themselves) with a dimension of 2 studs, as shown in Figure 9-8. This cube could be oriented with the studs pointing in any of the six possible directions and still fit within a space that's 2 studs wide, 2 studs deep, and 2 studs tall (again, disregarding the space occupied by the studs themselves).

Figure 9-8: A cube made from 20 1×1 plates, oriented along the three axes

HOW IT WORKS

To give you an appreciation of the logic involved, here's a conceptual description of how LSculpt works under the hood. Say you want to create a sculpture that's 16 studs wide. The software could break the 3D mesh down into a voxel grid that's 8 voxels wide. Each voxel would have a dimension equivalent to 2 studs, the same as one of the cubes shown in Figure 9-8. Let's call this the coarse voxel grid. In parallel, the software can also scale the 3D model by a factor of 5/2 and voxelize it into plates, similar to creating a studs-up sculpture. This time, however, there would be three different voxelized representations to account for three different orientations of the plates' studs: one with the 5/2 scaling done along the x-axis, one along the y-axis, and one along the z-axis. These three voxelized representations (let's call them A, B, and C) would also be aligned with the coarse voxel grid created earlier, such that each coarse voxel can contain 20 or fewer plates oriented along one of the three axes (as appropriate to capture the finer details of the model's shape).

The studs-out sculpture can be created by going through each coarse voxel and picking the orientation of plates (from A, B, or C) that would preserve the most amount of detail from the original model. What's the best way to determine this orientation? The algorithm LSculpt uses considers the surface normals of the triangles in the 3D mesh that are contained within the space occupied by each coarse voxel and figures out the dominant orientation. Given that A, B, and C are aligned to the same coarse voxel grid, we won't have any gaps or overlaps in the resulting studs-out sculpture.

MANUAL INTERVENTION

Because LSculpt outputs the plan for the sculpture entirely using 1×1 plates in different panels, the final step is to manually consolidate the sculpture into larger pieces that form an interlocking structure for each of the panels. For example, Figure 9-9 shows the different panels making up a studs-outs heart sculpture, complete with the SNOT core needed to hold everything together.

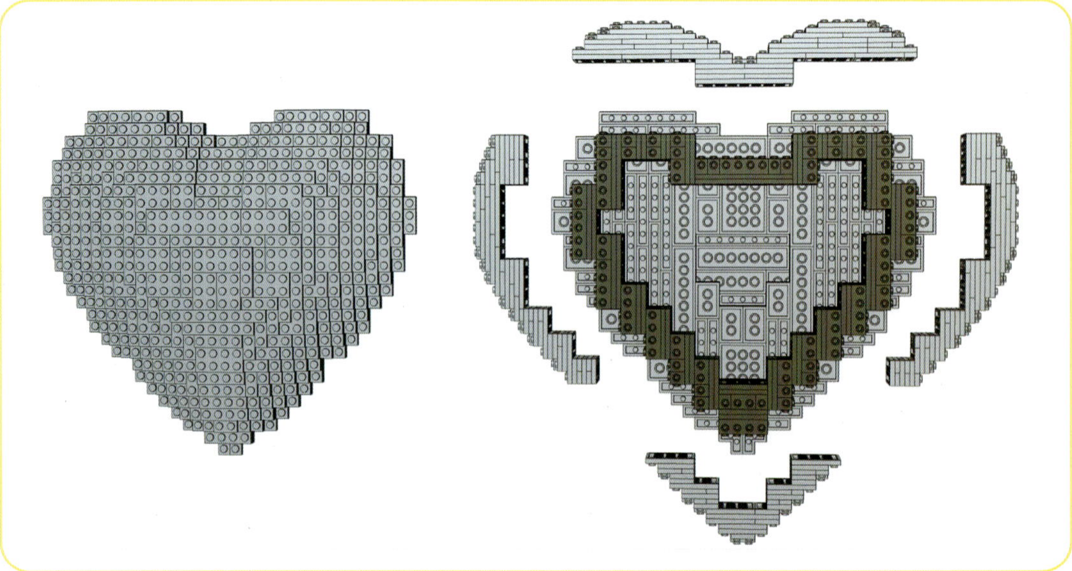

Figure 9-9: The six panels that make up the studs-out sculpture of the heart

DIGITAL BUILDING TIP

Consolidating 1×1 plates into bigger pieces can be tedious in Studio. The Merge tool can speed up the process. Select multiple adjacent plates by holding down CTRL as you click them, then right-click one of them and choose **Merge**. This replaces the selected pieces with a single larger piece, as long as a piece of the right dimensions exists in the LEGO catalog. If you aren't happy with a merge and want to try a different configuration, right-click a piece and choose **Split** to break it back up into 1×1 plates.

Besides planning the interlocking structure, you may need to manually adjust the orientation of plates in some parts of the sculpture to ensure that it can be built using real pieces. This is because the algorithm for assigning the plate orientations doesn't always work well, especially for complex 3D models that have features thinner than the size of each coarse voxel. In this case, the surface normals for the triangles within each coarse voxel may be pointing in different (maybe even opposite) directions, making it harder to determine a single dominant orientation. LSculpt has ways to work around this situation, but the result may not always be perfect.

SUMMARY

In this chapter, we've covered the basic types of LEGO sculptures along with a broad overview of the steps involved in their creation. LEGO sculptures continue to be among the most challenging builds for AFOLs, given that the process is nowhere close to being fully automated, even with the use of software. This concludes the main chapters of the book, but see the next section for suggestions on where to go from here.

NEXT STEPS

It's impossible for one book to catalog every single building technique, but I hope the techniques we've discussed have given you a good starting point for your own exploration of the LEGO medium. But where do you go from here, and how do you apply the techniques you've learned in this book to your own builds?

You'll probably have lots of questions as you start thinking about your next steps. Here are some answers to help guide you on your LEGO journey.

WHAT'S THE BEST WAY TO DESIGN A MOC?

Some people have an impromptu approach, playing around with a random pile of pieces and coming up with a design through trial and error. This, of course, requires having an inventory of LEGO pieces on hand. Others prefer a more deliberate, planned approach that involves measurements and calculations (which can be important if you're designing a replica of a real-life object) and maybe even sketches on paper. Software like BrickLink Studio can eliminate the need for paper sketches and help you quickly figure out how to lay out the pieces in your model. The bottom line is that there's no right or wrong way to design a MOC; it really depends on what you want to build and how you approach the creative process.

WHAT LEGO PIECES DO I NEED FOR MY MOC?

Going from official LEGO sets that have all the pieces already picked out to developing your own creation can be a little intimidating at first. LEGO makes thousands of different types of pieces. How do you know which ones are right for what you're trying to build?

This book has given you an overview of the basic types of pieces as well as some specialized ones (jumper plates, SNOT elements, hinges, and so on) that can be used for specific applications. You can get more familiar with the rest of LEGO's catalog of pieces by browsing through BrickLink. The entire catalog is also available as a building palette when you use software like BrickLink Studio.

If you have plans to turn your digital model into a physical one, you have to ensure that each piece you're using is actually made by LEGO in the color you need (Studio will warn you if you pick pieces in colors that aren't available). It also pays to keep an eye on the average cost of each piece you're using (this is also displayed in Studio) to avoid using rare, expensive pieces unless absolutely necessary.

WHERE CAN I BUY LEGO PIECES?

Before you buy any pieces to build a model, keep in mind that you can always repurpose parts from official LEGO sets that you already have. If you do need to stock up, though, LEGO stores usually have *Pick-a-Brick* walls with a wide variety of pieces that you can buy in bulk quantities. You can fill standard-sized containers with LEGO pieces of any kind and pay a flat price per container. The LEGO website also has an online version of Pick-a-Brick that allows you to order pieces in the quantities you need.

Another option is BrickLink, which offers the convenience of buying a much wider selection of LEGO pieces (and official sets) from a network of thousands of independent sellers around the world. The only tricky part is that each seller can have their own terms, shipping rates, and so on. You'll need to pay attention to those details before you place your order.

HOW DO I KEEP TRACK OF THE PIECES I NEED?

Here's an area where digital building tools really shine. If you're designing a model using a tool like Mecabricks or LEGO Art Remix, you have the option to save the list of required pieces in a BrickLink XML format. That file can then be uploaded to BrickLink, where you can order the pieces you need, saving you the trouble of manually entering each type of piece, its color, quantity, and so on. With BrickLink Studio, there's the added convenience of directly uploading your wish list (based on the design you've just created) to BrickLink.

WHAT'S THE BEST WAY TO SORT AND ORGANIZE THE LEGO PIECES IN MY COLLECTION?

As your LEGO collection grows, it becomes more important to keep it well organized. People typically sort their LEGO pieces based on type, color, or a combination of both. There's no right or wrong way to do it; it all comes down to the number and types of pieces you have and whatever you feel makes it easiest to quickly find that one piece you need.

There are several resources available that discuss this topic. A particularly useful and comprehensive one is the online book *The LEGO Storage Guide* by Tom Alphin, freely available at *https://brickarchitect.com/guide/*. The book delves into all aspects of LEGO organization as well as the various storage solutions available to builders with small, medium, or large collections of LEGO pieces. Alphin's website also offers downloadable labels for nearly 1,500 different types of pieces. You can print these labels and affix them to the storage containers you're using to make your collection easier to navigate.

IS IT BETTER TO BUILD DIGITALLY OR USING REAL PIECES?

Digital and physical building don't have to be mutually exclusive. They can coexist, with each style of building having its own place in the LEGO hobby. Building digitally can help you quickly try out different ideas without having to worry about the cost of buying LEGO pieces or having the space needed to assemble and store your models. You can then choose to convert one or more of your digital designs into physical models built using real pieces.

Realizing a digital design with physical pieces may take way more work than simply reproducing the model brick for brick. Software tools like Studio have basic stability checks, but they aren't perfect at accounting for real-world effects like gravity that a physical model will encounter. Expect some trial and error, and be prepared to add structural reinforcement. This will ensure that your model is practical and sturdy enough for the real world.

But why use real pieces at all when you can easily create photorealistic renders of your digital builds? These renders may work great for social media posts, but don't go to a LEGO convention and expect people to huddle around a small screen to look at renders of your latest digital model. No matter how impressive your model is, sometimes there's no substitute for something built using real, physical pieces. Also, using your hands to sort LEGO pieces and build your models one piece at a time is an intrinsic part of the LEGO experience that simply can't be replicated in the digital realm.

WHERE CAN I FIND MORE INFORMATION ABOUT LEGO BUILDING TECHNIQUES?

The building techniques covered in this book only scratch the surface of everything that's possible using LEGO. There's so much more to explore, and you can do that using the wealth of other resources that are available in print and online. These include books (No Starch Press has an entire line of LEGO books), print magazines like *BrickJournal* (*https://brickjournal.com*) and *Blocks* (*https://www.blocksmag.com*), and websites like the Brothers Brick (*https://www .brothers-brick.com*), BrickNerd (*https://bricknerd.com*), Brickset (*https://brickset.com*), New Elementary (*https://www.newelementary.com*), Tips and Bricks (*https://www.tipsandbricks .co.uk*), and more. Don't forget another valuable resource—the instructions that came with the official LEGO sets that you already have. If you pay close attention to how these official sets are put together, you may be able to glean techniques from them that you can adapt to your own builds.

HOW CAN I BECOME A MORE ACTIVE MEMBER OF THE LEGO COMMUNITY?

There are many ways to plug into the greater LEGO community, both in person and online. Keep an eye out for any LEGO conventions that may be held close to where you live. Attending these conventions can give you the opportunity to interact with other members of the LEGO community and check out all the cool LEGO models on display. This may even inspire you to think about displaying some of your own creations at a LEGO convention.

There may also be a *LEGO user group (LUG)* in your area, a group made up of local LEGO fans who meet on a regular basis and share their ideas and creations. LUGs sometimes also engage in collaborative group builds that are displayed at LEGO conventions. There's no reason to be intimidated about joining a LUG, even if you have little or no LEGO building experience. Most LUGs are happy to welcome new members, no matter their ability level. Being a part of a LUG is a great way to learn and level up your skills.

Even if you don't have any LEGO conventions or LUGs nearby, you can always use the internet and social media platforms to interact with fellow LEGO fans from around the world. Deserving special mention are Eurobricks (a long-running forum where members share their creations and discuss everything LEGO-related) and the Flickr and Facebook groups devoted to every facet of the hobby you can think of (castle builds, Star Wars–themed builds, mosaics, and more).

A great way to engage with the community and improve your LEGO skills at the same time is to participate in online LEGO-building contests. Some examples that recur every year are Summer Joust, SHIPtember, and FebRovery. Themed contests like these will challenge you to try out different kinds of builds and broaden your LEGO experience.

HAPPY BUILDING

We've covered a wide range of building techniques in this book, from basic ways of stacking LEGO bricks to methods for designing complex sculptures. Not every technique will be relevant to every build, and that's okay; you can always return to this book whenever you need a refresher or some new ideas. Let the book be a reference you can continue to consult while you grow as a builder.

With that, we'll wrap this book up. While this may be the end of our journey exploring LEGO building techniques together, hopefully it's just the start of a new chapter of the LEGO journey for you. As you work on making the most of what you've learned in this book, don't forget to also have some fun. In the end, that's what the LEGO hobby is all about. Happy building!

IMAGE CREDITS

INDEX